Dear Kathy & Ron,

    Thanks for your friendship and interest in native plants.

        Your friend
        Russ
        5 August 2006

Prairie Plants of the Midwest: Identification and Ecology

By Russell R. Kirt
College of DuPage

Illustrations by Henrietta H. Tweedie and Roberta L. Simonds

ISBN 0-87563-573-3

Copyright © 1995
Stipes Publishing L.L.C.

Published by
STIPES PUBLISHING L.L.C.
10–12 Chester Street
Champaign, Illinois 61820

Cover illustration by Henrietta H. Tweedie.

Photo of author and artists by Gene Sladek at West Prairie-Marsh Nature Preserve, College of DuPage, Glen Ellyn, IL.

Photo of Prairie Cordgrass, *Spartina pectinata,* by Russell R. Kirt at West Chicago Prairie, West Chicago, IL.

No illustrations in this book may be used without permission in writing from the illustrator(s).

ACKNOWLEDGEMENTS

Encouragement for writing this book came from many persons: notably my wife, Pamela. She has also been a constant partner in the planning, preparation, and editing of this book. Her ideas and suggestions have been incorporated and much of the quality in this book is due to Pamela's involvement.

Much enthusiasm and all the illustrations were provided by Henrietta H. Tweedie and Roberta L. Simonds. The work of these two exceptionally gifted artists give *Prairie Plants of the Midwest: Identification and Ecology* the visual dimension needed to accompany plant descriptions.

One of my former students, Thomas Simpson, Ph.D., of The Morton Arboretum, helped to proofread and made valuable suggestions for the Prairie Background section. David Malek and Lynn Fancher from College of DuPage reviewed and proofread the manuscript. Their comments were helpful and appreciated.

My most sincere gratitude goes to Ray Schulenberg who not only taught and inspired me to restore prairie but also gave me information, advice, and constructive criticism in the preparation of this book. This book would not be possible without Ray's involvement. In addition to being my mentor, I am most grateful for the strong ecological ethics and attitudes he instilled in me. The founder of Schulenberg Prairie at The Morton Arboretum in Lisle, Illinois set the standards for most prairie restoration projects in the midwest.

This book is dedicated to Ray Schulenberg.

PREFACE

This book is written and illustrated to aid the novice in the identification of tallgrass prairie plants native to the midwest. In this book, the plants are arranged by family because of the prominence of three families in the prairie, that is, the grass family (Poaceae), the legume family (Fabaceae), and the composite family (Asteraceae). A plant family is a broad grouping of related genera that share common characteristics. The Latin name of a family usually ends in "aceae," for example, Liliaceae (the lily family). Most of the nomenclature is from Swink and Wilhelm, 1994; some is from Gleason and Cronquist, 1991. A phenological (flowering date) chart is included on pages xi–xv so that the observer will know which plants are in bloom at any particular time.

Illustrations are intended to be a major tool for identification in this book. Therefore, plant descriptions are brief and given with a minimum of technical terms. When technical terms are used they are defined in the glossary. Whenever two species are likely to be confused, more detailed differences are given using both plant description and additional illustration. Those who wish a more scientific approach to plant identification may consult a technical book, such as *Plants of the Chicago Region*, 4th Edition, by Swink and Wilhelm.

Ecological notes are presented for the three most prominent plant families, and for some genera and most species. These notes provide information about the general importance of the plants to the prairie community, plant pollination ecology, unique anatomical structures enabling the species to survive in the prairie community, and other points of interest.

PLANT NAMES

There is a human need to name plants. Once a person correctly learns the name and identification of a prairie plant, he/she is likely to remember some of its ecology and importance to the prairie community. Most plants have a common and a scientific name which are usually complementary and have definite meanings.

Common Names

The common name is often locally recognized and consists of everyday words; however, it is seldom universal. Unfortunately many plants have the same or similar common names, and in some cases a plant may have several common names. In this book, prairie plants are designated by the common names most frequently used in the midwest.

Scientific Names

Each plant has only one valid scientific name, that is, the species. The scientific name is made up of three or more parts: (a) the genus, (b) the specific epithet, and (c) the authority who named the plant (Note: The author's name is often abbreviated). The genus (plural: genera) is a group of closely related species. The specific epithet is often referred to as the "species" but this is incorrect. The genus name and the specific epithet together form the name of the species. The word "species" is both singular and plural. A species is a group of genetically and structurally related organisms which are capable of producing fertile offspring. An example of a scientific name is *Dodecatheon meadia* L. (the common name for shooting star). The genus is *Dodecatheon*, the specific epithet is *meadia*, and the author is L. (abbreviation for Linnaeus).

Sometimes a plant species has a recognizable subspecies or variety which is included. For example, *Phlox glaberrima interior* Wherry is the scientific name for marsh phlox. The variety name for this plant is *interior*. The genus, specific epithet, and variety should be underlined or italicized in written literature.

For good botanical/taxonomic reasons, plant names are occasionally changed to another specific epithet or genus following rules set forth by the International Code of Botanical Nomenclature. An example of a plant whose name has been changed is Culver's root, *Veronicastrum virginicum* (L.) Farw. The name in parentheses is that of the person(s) who first applied the scientific name, followed by the person(s) who transferred the plant to another genus or gave it a different specific epithet.

It is sometimes not necessary or possible to give the specific epithet. In these cases the abbreviations, sp. (singular) and spp. (plural), are used. The abbreviations are not underlined or italicized. In this book the abbreviation "spp." is often used when referring to more than one species belonging to the same genus. For example, the indigos are discussed as *Baptisia* spp.

PRAIRIE BACKGROUND

Over the past 12,000 to 15,000 years in the mid-continent of North America, forces of climate, fire, and grazing created conditions that favored the development of grassland. This biome is generally known as prairie east of the 100th meridian and as great plains to the west of the 100th meridian. In North America, most plant ecologists divide this biome into tallgrass prairie, midgrass prairie, and shortgrass prairie.

NORTH AMERICA GRASSLANDS

Adapted from A. W. Küchler.

This book will focus primarily on the plants of the tallgrass prairie, an ecosystem that once covered almost 264,000 square miles (68,378,112 hectares) from Canada to Texas and Nebraska to the Great Lakes. Most of the tallgrass prairie vanished in less than 50 years, converted into farmland by the plow and overgrazed by domestic livestock. Today only small patches of tallgrass prairie remain, barely enough to remind us of its original grandeur. There is no tallgrass prairie national park; this ecosystem disappeared too quickly. There were only a few conservationists interested in prairie preservation prior to the 1970's. As a result, pristine tallgrass prairie is the rarest of North America's major biomes. Approximately 99.99% of the tallgrass prairie has been destroyed; some of the remaining 0.01% is protected. Two of the largest tallgrass prairie preserves are the Konza Prairie in northeastern Kansas and the Tallgrass Prairie Preserve in north central Oklahoma; both of these prairies are owned by The Nature Conservancy. Even today, non-protected prairie remnants are threatened by housing and shopping center developments, and paving for roads and parking lots.

In the power of the sun and wind, the vast greenness of treeless tracts and spacious skies, and the changing seasonal moods of color and texture, prairie has no equal in North American landscape. In tallgrass prairie, plant life is rich and diverse. Seas of grass grow as tall, or taller, than a man's head. Spectacular flower blooms begin with pasque flower and prairie smoke in early spring, followed by coneflowers, blazing stars, and sunflowers during summer, and end with asters and gentians in fall. No other ecosystem produces this continuous spectacle of wildflower blooms throughout the growing season.

Geologic history, climate, fire, and animals have all shared a role in the origin and maintenance of prairie.

Prairie rests on a firm foundation. A great plain of sedimentary rock underlies central North America. It is formed from shells, sand, and silt of inland seas that periodically flooded the mid-continent during the Paleozoic Era, between 600 million and 280 million years ago. This sedimentary platform of limestone, sandstone, and shale now lies hundreds of feet above sea level, but remains relatively flat.

Glaciers first crept southward from the frigid subarctic about two million years ago, at one time advancing as far south as the southern tip of Illinois and northern Missouri. Glaciers eroded the soft sedimentary rock and further flattened the landscape. On top of this flattened plain, glaciers dumped their load of ground up bedrock. This mass of sand, silt, clay, pebbles, and boulders covers glaciated terrain, often hundreds of feet thick. On top of this glacial till is a layer of silt, called "loess," blown by Ice Age winds from the glacial river valleys. Loess deposits are over 300 feet thick in western Iowa near the Missouri River, yet only a few inches thick in areas such as the Flint Hills of Kansas and Oklahoma.

Moist air from the Pacific Ocean rises and cools as it ascends the windward side of the western mountains. Precipitation falls heavily on the western slopes, leaving only dry air to cross the continental divide. The mountains cast a long rain shadow over the grasslands to the east.

Grassland rainfall generally varies from as little as ten to thirty-five inches per year. In North America, grasslands just east of the Rocky Mountains receive an average of sixteen inches of rainfall per year. In this region, known as shortgrass prairie, grass height varies from about six to eighteen inches. Eventually, the rain shadow is alleviated somewhat by masses of moist air moving northward from the Gulf of Mexico. In western Kansas and Nebraska, average rainfall increases to about twenty-five inches per year; here midgrasses grow to a height of two to four feet. From central Nebraska to Indiana, the average rainfall increases to about thirty-three inches per year. In this tallgrass prairie, grasses grow to a height of six to seven feet. Precipitation ranging from ten to thirty-five inches per year, along with hot dry summers and cold winters, is the climate needed for grasslands. A grassland climate typically includes droughts, sometimes lasting a decade.

Where the average rainfall approaches, or is greater than, thirty-five inches per year, the tallgrass prairie often blends into open forest. This mixture of prairie and forest is generally known as "savanna." In savanna, the tree canopy does not totally shade the ground, therefore herbaceous plants have sunlight to photosynthesize for at least part of the day. In the midwest, especially in Illinois, savannas were referred to as "groves." Oaks are the dominant tree species of savanna. In sandy soil, black oak, *Quercus velutina* Lam., is the dominant tree species. Where mesic soil conditions exist, bur oak, *Quercus macrocarpa* Michx. is the major tree species. Bur oak has a thick bark which makes the tree resistant to fire.

SAVANNA

Fire is a major abiotic factor in maintaining prairie ecosystems. In presettlement times, fires occasionally burned over the prairie during fall or early spring when the plants were dormant and their stems and leaves were dead and dry. They were caused by lighting and by Native Americans who routinely set fire to the prairie to attract large game animals. Buffalo, elk, and deer would graze on the profusion of succulent tender green plants that grew after fire. These animals were subsequently hunted for food. Sometimes in fall, fires were started to harvest large numbers of buffalo and other game mammals by driving them into an area where they could be easily hunted.

Most prairie plants are not harmed by fire. Roots and other underground organs comprise approximately two-thirds of most prairie plants. Each year, herbaceous stems grow from buds, rhizomes, and other plant tissue located beneath the soil's surface and safe from the fire's heat. This subterranean adaptation to fire gives prairie plants an advantage over trees and shrubs. Without fire, the prairie is weakened by the accumulation of unburned litter. The litter provides a cooler and shadier microclimate, a disadvantage to most prairie plants but an advantage to woody species.

Fire releases nutrients from the ash of burned mulch more quickly than decay. Recently burned prairies produce about twice the annual biomass of those not burned for several years. Fire also helps eliminate weeds; plants not part of a stable prairie ecosystem.

Buffalo herds numbering in the thousands are no longer available to view. From 1870 to 1885, nearly 60 million bison that roamed the prairie and plains of North America were exterminated. By 1900, fewer than one thousand remained. Today there may be 150,000 surviving bison, thanks to herds maintained by The Nature Conservancy and private individuals. Buffalo concentrate grazing on tender new grasses and move on when these grasses are gone. The bison did little overall injury to the prairie. Their intermittent grazing left enough dry, dead grass and forbs behind to fuel the next prairie fire.

The smallest and most numerous fauna in prairie are the insects. Wasps and bumblebees, followed by butterflies and moths, are the major pollinators of flowers. Some butterflies and other insects need specific prairie plants for food and reproduction; in return, the insects pollinate the flower. Insects, therefore serve as sensitive ecological indicators, revealing the health of the prairie in which they live.

The prairie was a delicate balance between flora and fauna. Populations of burrowing mammals, such as prairie dogs and ground squirrels, are estimated to be reduced by nearly 98%. A variety of predator animals, such as fox, coyote, hawks, and snakes were supported by these burrowing mammals. Less obvious are the ecological changes resulting from the decreased role of burrowing animals in nutrient cycling and soil formation. Population declines for grassland birds vary from 24 to 91%. Of the 435 bird species breeding in the United States, 330 breed in the grasslands. Many species, such as the Henslow's sparrow, Attwater's prairie chicken, upland sandpiper, and short-eared owl, are extremely rare, endangered, or threatened today.

With the loss of the tallgrass prairie, we have lost much more than grasses and forbs. We have lost an entire ecosystem. Bison, prairie dog, falcon, and compass plant all shared the same fate, together. The fragments of prairie that are with us today are a part of our heritage, a priceless reminder of a vast and beautiful landscape.

PHENOLOGICAL (FLOWERING DATE) CHART

The solid line indicates the average blooming dates of the prairie plant species in this book. This chart does not include the earliest flowering or latest flowering dates during years having a very early spring or a very late autumn.

	APRIL	MAY	JUNE	JULY	AUG	SEPT	OCT
Allium cernuum, Nodding Wild Onion					——		
Amorpha canescens, Lead Plant				——			
Andropogon gerardii, Big Bluestem						——	
Andropogon scoparius, Little Bluestem						——	
Anemone canadensis, Meadow Anemone			——————				
Anemone cylindrica, Thimbleweed			——————				
Anemone patens, Pasque Flower	——————						
Asclepias hirtella, Tall Green Milkweed				——————			
Asclepias sullivantii, Prairie Milkweed				——————			
Asclepias tuberosa, Butterfly Milkweed			——————				
Aster azureus, Sky Blue Aster						———————————	
Aster ericoides, Heath Aster					——————		
Aster laevis, Smooth Blue Aster					——————————————		
Aster novae-angliae, New England Aster					——————————————		
Aster sericeus, Silky Aster					——————————————		
Astragalus canadensis, Canadian Milk Vetch				——————			
Baptisia leucantha, White Wild Indigo			——————				
Baptisia leucophaea, Cream Wild Indigo		——————					
Bouteloua curtipendula, Side-oats Grama					——————		
Bromus kalmii, Prairie Brome			——————				
Cacalia plantaginea, Indian Plantain			——————————				
Calamagrostis canadensis, Blue Joint Grass			——————————				
Camassia scilloides, Wild Hyacinth		——————					

	APRIL	MAY	JUNE	JULY	AUG	SEPT	OCT
Carex bicknellii, Prairie Sedge		——					
Carex meadii, Mead's Sedge		—					
Carex pensylvanica, Pennsylvania Sedge	——						
Carex stricta, Strict Sedge		——					
Castilleja coccinea, Indian Paint Brush		——————					
Ceanothus americanus, New Jersey Tea			——————				
Cirsium discolor, Pasture Thistle					————————		
Comandra umbellata, False Toadflax		————					
Coreopsis palmata, Prairie Coreopsis			——————				
Coreopsis tripteris, Tall Coreopsis					————————		
Cypripedium candidum, White Lady's Slipper		——					
Desmodium canadense, Showy Tick Trefoil				————————			
Desmodium illinoense, Illinois Tick Trefoil				————————			
Dodecatheon meadia, Shooting Star		——					
Echinacea pallida, Purple Coneflower				——			
Elymus canadensis, Canada Wild Rye				————————			
Eryngium yuccifolium, Rattlesnake Master				————————			
Euphorbia corollata, Flowering Spurge				————————			
Fragaria virginiana, Wild Strawberry	——————————						
Galium boreale, Northern Bedstraw			——————				
Gentiana andrewsii, Bottle Gentian						————————	
Gentiana flavida, Yellowish Gentian					————————		
Gentiana puberulenta, Downy Gentian						————————	
Geum triflorum, Prairie Smoke		——					
Habenaria leucophaea, Prairie White Fringed Orchid				——			
Helianthus mollis, Downy Sunflower					————————		

	APRIL	MAY	JUNE	JULY	AUG	SEPT	OCT
Helianthus occidentalis, Western Sunflower					———	————	
Helianthus rigidus, Prairie Sunflower						———	———	
Heliopsis helianthoides, False Sunflower				——	————	———	
Heuchera richardsonii, Alum Root			—	—			
Hierochloe odorata, Vanilla Grass			—				
Hypoxis hirsuta, Yellow Star Grass			—				
Iris virginica shrevei, Blue Flag Iris			—	—			
Koeleria cristata, June Grass				—			
Kuhnia eupatorioides, False Boneset						—	—	
Lespedeza capitata, Round-headed Bush Clover					|—|	
Liatris aspera, Rough Blazing Star						———	———	
Liatris pycnostachya, Prairie Blazing Star				—	—		
Lilium michiganense, Turk's Cap Lily				—			
Lilium philadelphicum andinum, Prairie Lily				—			
Linum sulcatum, Grooved Yellow Flax					—	—	
Lithospermum canescens, Prairie Puccoon			—————	————			
Lobelia spicata, Pale Spiked Lobelia				—	—		
Monarda fistulosa, Wild Bergamot					—	—	
Oenothera pilosella, Prairie Sundrops				—	—		
Opuntia humifusa, Prickly Pear				—	—		
Oxalis violacea, Violet Wood Sorrel			—				
Panicum oligosanthes scribnerianum, Scribner's Panic Grass				—			
Panicum virgatum, Switch Grass						—	—	
Parthenium integrifolium, Wild Quinine				—————	————		
Pedicularis canadensis, Wood Betony		—	—				
Penstemon digitalis, Foxglove Beard Tongue				—			

	APRIL	MAY	JUNE	JULY	AUG	SEPT	OCT
Petalostemum candidum, White Prairie Clover				——			
Petalostemum purpureum, Purple Prairie Clover				——			
Phlox glaberrima interior, Marsh Phlox				—————			
Phlox pilosa, Prairie Phlox		——					
Physostegia virginiana, False Dragonhead				—————			
Polygala senega, Seneca Snakeroot		——					
Polytaenia nuttallii, Prairie Parsley		——					
Potentilla arguta, Prairie Cinquefoil			————				
Prenanthes aspera, Rough White Lettuce					———	———	
Psoralea tenuiflora, Scurfy Pea			——				
Pycnanthemum virginianum, Common Mountain Mint				————			
Ratibida pinnata, Yellow Coneflower				————————			
Rosa carolina, Pasture Rose			————				
Rudbeckia hirta, Black-eyed Susan			————————————				
Salix humilis, Prairie Willow	———						
Senecio pauperculus, Balsam Ragwort		———					
Silphium integrifolium, Rosin Weed				————	———		
Silphium laciniatum, Compass Plant				————	———		
Silphium perfoliatum, Cup Plant				————————			
Silphium terebinthinaceum, Prairie Dock				————————————			
Sisyrinchium albidum, Blue-eyed Grass		——					
Solidago missouriensis fasciculata, Missouri Goldenrod					———		
Solidago nemoralis, Old-field Goldenrod						———	
Solidago rigida, Rigid Goldenrod					———————		
Sorghastrum nutans, Indian Grass					———		
Spartina pectinata, Prairie Cord Grass					———————		
Sporobolus heterolepis, Prairie Dropseed					———		

	APRIL	MAY	JUNE	JULY	AUG	SEPT	OCT
Stipa spartea, Porcupine Grass			—				
Tradescantia ohiensis, Spiderwort		————————					
Vernonia fasciculata, Common Ironweed					———————		
Veronicastrum virginicum, Culver's Root				————			
Vicia americana, American Vetch		——					
Viola pedatifida, Prairie Violet		—					
Zizia aptera, Heart-leaved Meadow Parsnip		———					
Zizia aurea, Golden Alexanders		———					

APIACEAE (CARROT or PARSLEY)

Eryngium yuccifolium Michx., RATTLESNAKE MASTER

Height: 1–1.6 meters (3–5 feet)
Flowers: July–August
Color: White
Habitat: Mesic–Xeric

Identification Features:

The solitary stem of Rattlesnake Master grows from a short, thick rootstock. Thick, bayonet-shaped, alternate leaves clasp the stem. The leaves are parallel veined and have weak yucca-like bristles spaced far apart along the edges. The flower heads are white, hard, stiff, and prickly. This elegant plant is easily recognized throughout the growing season.

Ecological Notes:

Rattlesnake Master is a member of the stable prairie community. It decreases in prairie where ungulate grazing occurs as its new growth is palatable and nutritious. The flowers attract many species of small insects for pollination activities. The leaves are rarely eaten by insects, such as grasshoppers and caterpillars, due to the strong cords of lignified cells running along the margins and veins. Anatomical features, such as waxy leaves and sunken pores in the leaf's upper epidermis, may help reduce the loss of water vapor from Rattlesnake Master during drought conditions.

Polytaenia nuttallii, DC., PRAIRIE PARSLEY

Height: 0.6–1.0 meter (2–3 feet)
Flowers: June
Color: Yellow
Habitat: Xeric-Mesic

Identification Features:

Prairie Parsley has an erect, stout, finely pubescent to smooth stem that grows from a long thick taproot. The lower leaves have long petioles and are deeply divided into narrow segments. The upper leaves are opposite and also deeply divided. Small, yellow flowers are borne in six to twelve flat-topped umbels at the ends of upper branches. The fruits are oblong to elliptic, flattened, 6–8 mm long, and yellow-green.

Umbel

Ecological Notes:

Prairie Parsley is one of the few biennials of the prairie. The great majority of prairie plants are perennials.

Biennials normally form a rosette the first year; during the second year, they send up a flowering stalk, bear seed, and die. However, Prairie Parsley and some other plants which are normally biennial may live as rosettes for more than one year.

Zizia aptera (Gray) Fern., HEART-LEAVED MEADOW PARSNIP

Height:	0.3–0.6 meter (1–2 feet)
Flowers:	Mid-May to Mid-June
Color:	Yellow
Habitat:	Mesic-Xeric

Identification Features:

Heart-leaved Meadow Parsnip has erect, smooth stems which grow from a cluster of thickened roots. Its basal leaves are on long petioles, shallowly toothed, and "heart-shaped." The upper leaves are divided into three leaflets. The leaflets are ovate to lancelate and finely toothed. The bright yellow flowers are small and are borne in compound umbels. The fruits are oblong, smooth, 3 mm long, and have narrow ribs.

Ecological Notes:

Heart-leaved Meadow Parsnip is a member of stable prairie and savanna communities.

Basal Leaf ⟶

Zizia aurea (L.) W. D. J. Koch, GOLDEN ALEXANDERS

Height: 0.3–1 meter (1–3 feet)
Flowers: May–Mid June
Color: Golden-Yellow
Habitat: Hydric-Mesic

Identification Features:

Golden Alexanders has one to eight erect stems that are branched. Its basal and lower leaves are on long petioles; they are divided in threes, sometimes twice. The upper leaves are on short petioles and are sometimes divided into three leaflets. The leaves are irregularly shaped, bright green, and have fine sawtoothed margins. The golden-yellow flowers are small and occur in compound umbels. The whole umbel is like a bouquet.

Ecological Notes:

Golden Alexanders thrives in moist prairies and in savanna.

ASCLEPIADACEAE (MILKWEED)

Ecological notes and pollination mechanism for the Milkweeds (*Asclepias* spp.):

Milkweed flowers are arranged in rounded or flat clusters called umbels. Each flower is composed of a corona with five petal-like hoods, five nectar horns with keyhole-like niches between them, and a corolla of five recurved petals which often conceals the calyx. Pollen sacs (pollinia), which are attached to a Y-shaped structure within the corona, are located in slits between the anthers. When an insect comes to gather nectar, it invariably slips its legs into a niche. The insect is held firmly while it sips nectar but as it struggles to free itself, attached pollinia collected from previously visited milkweeds are sloughed off and other pollinia are picked up. Usually the insect is able to free itself by pulling its foot upward and out. However, some insects are unable to escape and die on the flower.

The above described mechanism ensures cross-fertilization for the milkweed flowers. Although there are dozens of flowers in the cluster, usually only two or three become properly pollinated. The pollinated flowers form large downy pods full of silk-plumed seeds.

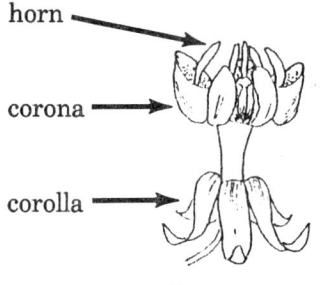

Milkweed Flower

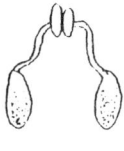

Pollinia

Asclepias hirtella (Pennell) Woodson, TALL GREEN MILKWEED

Height: 0.3–1.0 meter (1–3 feet)
Flowers: July–Mid August
Color: Greenish-Yellow
Habitat: Hydric-Mesic

Identification Features:

The smooth stem of Tall Green Milkweed grows from a root system that has numerous thread-like roots. The linear to lanceolate leaves are mainly alternate and rough to the touch. Both the stem and leaves contain a milky latex. The flowers are borne in umbels from the leaf axils toward the top of the stem. The erect pods contain an average of 50 plumed seeds.

Ecological Notes:

Tall Green Milkweed is a characteristic species of quality hydric sand prairie communities. Individual plants may have ripened pods that are shedding seeds at the base of the plant, while at the upper end of the stem the umbels are still being visited by insects. This flowering and seed maturing process enables a large number of different insect species, such as bumblebees, to pollinate the flowers of Tall Green Milkweed.

Asclepias sullivantii Engelm., PRAIRIE MILKWEED

Height: 0.7–1.5 meters (2–5 feet)
Flowers: Late June–July
Color: Pink-Rose-Purple
Habitat: Hydric-Mesic

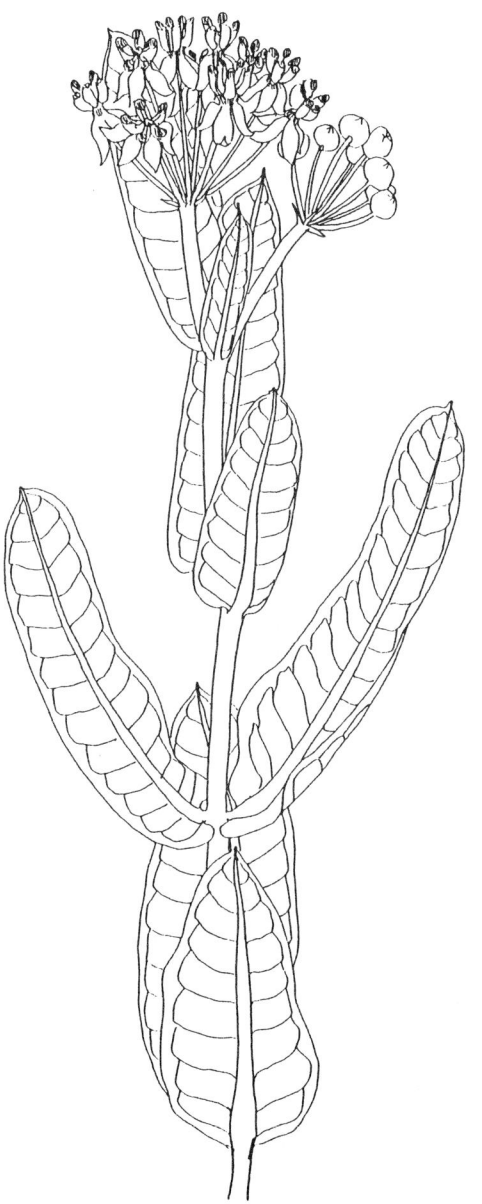

Identification Features:

The smooth, stout stem of Prairie Milkweed is pink to green-white in color. Its thick leaves are sessile, hairless, and point upward. They have conspicuous pink veins in addition to a strong, pink midrib. The umbels are terminal and wide-spreading. The individual blossoms are rose-pink, 1 cm in diameter, and larger than those of the Common Milkweed.

Ecological Notes:

The presence of Prairie Milkweed often indicates virgin prairie. As with all milkweeds, its latex contains bitter alkaline chemicals which usually give the mature milkweed protection from most leaf and stem herbivores. There is a noticeable absence of insect holes in the stems and leaves of all the milkweeds. However, one of the few leaf and stem herbivores known to eat the acrid leaves of the Prairie and Common Milkweed is the larva of the monarch butterfly. The alkaline substance accumulates in the larva, and gives the adult butterfly some protection from bird predators.

Asclepias tuberosa L., BUTTERFLY MILKWEED (PLEURISY ROOT)

Height: 0.3–0.6 meter (1–2 feet)
Flowers: Late June–Mid August
Color: Orange-Yellowish
Habitat: Xeric-Mesic

Identification Features:

Butterfly Milkweed has one to several erect hairy stems that arise from a stout tuberous taproot. Its alternate leaves are lance-like to tapered and are strongly pubescent. Butterfly Milkweed has three to nine orange-yellowish flowers arranged in one to several umbels for each of its flat-topped inflorescences. Butterfly Milkweed differs from other milkweeds in that its stems and leaves do not excrete a white latex when cut. This species is one of the most spectacular prairie plants in bloom.

Ecological Notes:

Butterfly Milkweed thrives along roadsides (especially where there is well-drained sandy soil), in savanna, and in prairie. Butterfly Milkweed is usually pollinated by flying insects as its hairy stems make it difficult for walking insects to climb up to the flowers. Ungulates do not eat Butterfly Milkweed; thus it tends to increase in degraded rangeland.

ASTERACEAE (COMPOSITE or SUNFLOWER)

General importance of the composite family to the prairie and other notes:

The composites have more species in the prairie flora than any other family. Members of this family contribute the most growth forms and beauty to the prairie community. Each stage of the growing season has a rich spectacle of composites in bloom.

The composites contribute to the soil building process. Some plants, such as Compass Plant, have thick taproots that penetrate deeply into the clay subsoil and help break it up. The rhizomes of composites, such as Asters, help bind the soil especially if it has been disturbed.

Many insects gather pollen from the showy composites. Some composites, such as the sunflowers, have large seeds that are eaten by grassland birds and small mammals. The thick roots of many composite members provide food for some small mammals during the late fall and winter months.

The composite head consists of two types of flowers: disk flowers and ray flowers. The smallest and most numerous flowers in the head are disk flowers which are generally located in the center. The strap-like "petals" surrounding the disk flowers are ray flowers. Many composites have both disk and ray flowers, while some have only one or the other.

Ecological notes for the asters, *Aster* spp.

One of the largest genera in the prairie community is *Aster*. Members of this genus occupy a wide variety of habitats. For example, New England Aster and Heath Aster tend to be early colonizing species and also thrive in degraded prairie. Other asters, such as Silky Aster, Sky-blue Aster, and Smooth Blue Aster, are members of the stable prairie community.

During fall, asters are a very important nectar source for a variety of insects, especially the butterflies.

Aster azureus Lindl., SKY-BLUE ASTER

Height: 0.2–1.2 meters (0.7–4 feet)
Flowers: Late August–October
Color: Blue-Purple rays, Yellow disks
Habitat: Xeric-Mesic

Identification Features:

Sky-blue Aster has erect and rigid stems which are widely branched toward the top. Scattered clumps of this aster grow from fibrous roots and from the bases of old stems. The basal and lower leaves have long slender petioles. Their blades are cordate to ovate, 2–5 cm wide at the base and 5–15 cm long. The upper leaves are almost sessile. They are lanceolate and shallowly serrated. Both the upper and lower leaves are thick and firm. Sky-blue Aster has an open inflorescence with sparse, long slender branches; its heads have ten to twenty-five blue-purple ray flowers and many yellow disk flowers.

Ecological Notes:

In the genus *Aster,* Sky-blue Aster has the greatest fidelity to undisturbed prairie and savanna communities.

Aster ericoides L., HEATH ASTER

Height: 0.3–0.6 meter (1–2 feet)
Flowers: September–Early October
Color: White rays, Yellow disks
Habitat: Mesic-Xeric

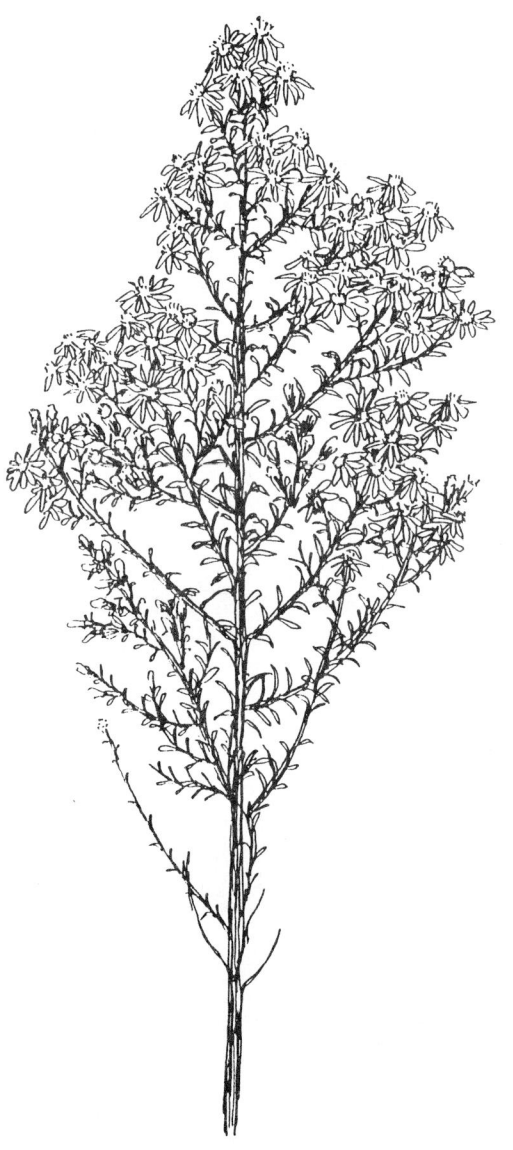

Identification Features:

Heath Aster usually grows in bush-like clumps from a root system which is highly rhizomatous. Its stem and leaves are covered with very short hairs. The numerous heads and leaves are tiny, especially toward the top of the plant, where they form a pyramid. The pyramid is 30–50 cm at the base. The composite head is about 5–6 mm in diameter and is composed of white ray flowers and yellowish disk flowers.

Ecological Notes:

Heath Aster is often abundant in prairies that are degraded due to factors such as overgrazing. Its abundance in degraded prairie is due to prolific seed production, low palatability, and extensive rhizome development. The dense rhizomatous root system helps bind the soil.

Aster laevis L., SMOOTH BLUE ASTER

Height: 0.3–1 meter (1–3 feet)
Flowers: Mid August–Mid October
Color: Blue-Violet rays, Yellow disks
Habitat: Mesic-Xeric

Identification Features:

Smooth Blue Aster has an erect stem that arises from a short rhizome. The alternate leaves are smooth, blue-green, and lanceolate. They have rough margins and clasp the stem. The basal leaves are much larger than those near the flowers. The flower heads are numerous with blue-violet ray flowers and yellow disk flowers.

Ecological Notes:

Smooth Blue Aster is a member of the stable prairie community.

Aster novae-angliae L., NEW ENGLAND ASTER

Height:	0.7–1.8 meters (2–6 feet)
Flowers:	August–October
Color:	Purple-Blue rays, Orange-Yellow disks
Habitat:	Hydric-Mesic

Identification Features:

The tall, leafy stalks of New England Aster protrude in all directions from its rhizomes. The lanceolate-shaped leaves are quite uniform in size and strongly clasp the stem. The terminal inflorescences are branched and have several flower heads. These large aster heads are up to 2.5 cm in diameter with purple-blue ray flowers and reddish-orange-yellow disk flowers. The flowers of New England Aster add much color to the prairie during autumn.

Ecological Notes:

New England Aster is an early colonizing species and declines as the prairie community matures. It can also survive severe disturbances.

Aster sericeus Vent., SILKY ASTER

Height: 0.1–0.6 meter (4–24 inches)
Flowers: September–October
Color: Lavender-Rose rays, Yellow disks
Habitat: Xeric

Identification Features:

Silky Aster has one to several erect and slender brownish-red stems. The stems are wiry and zig-zag near the middle. The small leaves are alternate, narrow, oblong to pointed, relatively uniform, and sessile. When the plant is flowering, the lower leaves have already fallen. Each head has fifteen to thirty lavender-rose ray flowers and many yellow disk flowers. Silky Aster has a distinctive appearance due to the silky, silvery hairs on both sides of the leaves.

Ecological Notes:

The presence of Silky Aster indicates high quality undisturbed prairie and savanna.

Cacalia plantaginea (Raf.) Shinners, INDIAN PLANTAIN

Height: 0.7–1.6 meters (2–5 feet)
Flowers: June–July
Color: Cream-White disks
Habitat: Mesic-Hydric

Identification Features:

Indian Plantain has an erect, smooth stem that arises from a small tuberous-thickened base and fleshy, fibrous roots. Its basal leaves are thick, oval, and have five to nine prominent parallel veins that converge toward the tip of the leaf blade. The basal leaves are often perforated with insect holes. The upper leaves are smaller and toothed near the tips. The flat-topped inflorescence is composed of white-cream flower heads with five tubular disk flowers. The achene has a very silky pappus.

For years, this plant was called *Cacalia tuberosa*, Nutt.

Ecological Notes:

The presence of Indian Plantain indicates virgin prairie conditions. This species grows best in calcareous soils.

Cirsium discolor (Muhl.) Spreng., PASTURE THISTLE

Height: 1-3 meters (3-8 feet)
Flowers: August-September
Color: Rose-Purple disks
Habitat: Mesic

Identification Features:

The conspicuous stem of Pasture Thistle arises from a deep taproot. The leaves are alternate, wavy, lanceolate, and coarsely toothed. Each of the coarse teeth on the leaf margin ends in a sharp spine. The underside of the leaf is silvery-white. The heads consist of rose-purple disk flowers surrounded by a green cup-like base of involucral bracts, each tipped with long bristles pointing outward. The seeds are flattened and oblong-shaped; each has a plume of soft bristles with white fluff on the underside.

Ecological Notes:

This thistle is generally a biennial (see Prairie Parsley, page 2, for description of the biennial life cycle). Pasture Thistle is a native of the prairie but survives in disturbed sites such as old fields. Bumblebees are the primary pollinators of thistles. Goldfinches are sometimes referred to as "thistle" birds because of their use of thistledown in nest building.

Cirsium hillii (Canby) Fern., PRAIRIE THISTLE, is very rare.

Coreopsis palmata Nutt., PRAIRIE COREOPSIS

Height: 0.6–0.9 meter (2–3 feet)
Flowers: Mid June–July
Color: Yellow-Orange rays, Brown disks
Habitat: Xeric-Mesic

Identification Features:

Prairie Coreopsis has a smooth, erect stems that arise from rhizomes. Its stiff leaves are opposite, sessile, palmately three-lobed (resembling a bird's foot) to the center, and about 5 cm long. The flower heads are up to 5 cm in diameter and contain both yellow ray and brown disk flowers. The ray flowers are often three-toothed at the tip. The foliage of Prairie Coreopsis turns to a beautiful orange-purple color during autumn.

Ecological Notes:

Prairie Coreopsis is highly rhizomatous and forms colonies that often exclude other species. Flowering usually occurs only on the perimeter of these colonies. Adult gorgone checkerspot butterflies rely heavily on Prairie Coreopsis as a host plant. Prairie Coreopsis grows in Black Oak savanna in addition to prairie.

Coreopsis tripteris L., TALL COREOPSIS

Height:	1–2 meters (3–7 feet)
Flowers:	August–September
Color:	Yellow rays, Brown disks
Habitat:	Mesic-Hydric

Identification Features:

The tall, smooth stem branches at the top and is capped by numerous flower clusters. The principle leaves are usually divided to the midrib into three leaflets. The yellow ray flowers are very showy. During fall, the leaves and stem of Tall Coreopsis turn a beautiful reddish-orange color.

Ecological Notes:

Tall Coreopsis grows under many conditions and is almost weedy. The flowers produce much pollen for bees and other pollinators at a time when there are not many forbs blooming in the prairie.

Echinacea pallida Nutt., PURPLE CONEFLOWER

Height: 0.5–0.9 meter (1.5–3 feet)
Flowers: Mid June–Early July
Color: Purple rays, Brownish-Black disks
Habitat: Xeric-Mesic

Identification Features:

Purple Coneflower usually has two to five unbranched pubescent stems that have a single terminal flower head. It has a deep, thick tuberous taproot, 1–3 cm in diameter. Its basal leaves are quite large, up to 18 cm long and 3–4 cm wide. They are ovate-lanceolate in shape, noticeably pubescent, and strongly three-veined. Smaller leaves ascend the stem. Large pinkish-purple flower heads extend up to 10 cm in diameter when first in bloom. However after approximately one week, the pinkish-purple ray flowers begin to droop and form an inverted "cone." After approximately three weeks, the ray flowers drop off and the central head of spiny disk flowers remain conspicuous until late autumn. This elegant plant is easily recognized throughout the growing season.

Ecological Notes:

Purple Coneflower is an indicator of high quality virgin prairie. This colorful forb is palatable and nutritious to ungulates. The roots are a preferred food source of voles and many other rodents. While it is in bloom, many insects seek out the yellow pollen from the disk flowers.

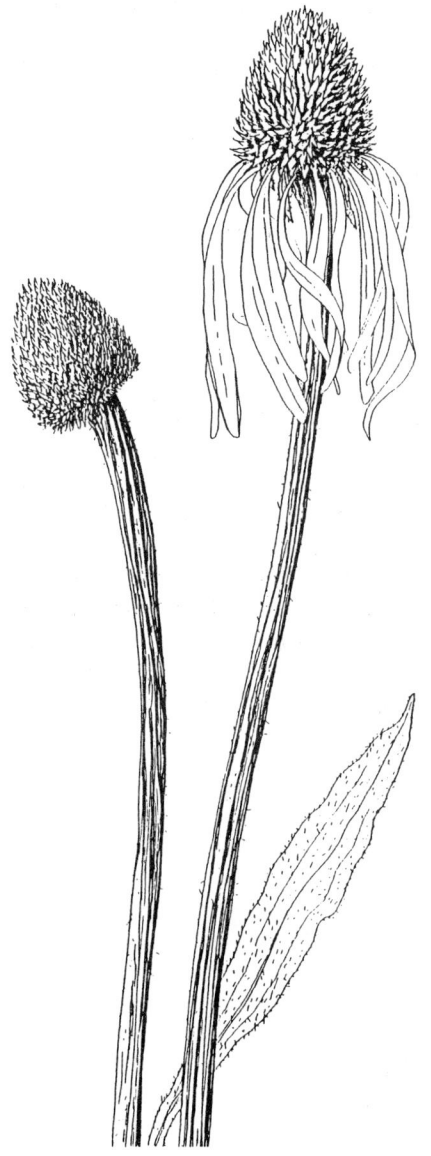

Ecological notes for the sunflowers, *Helianthus* spp.

Members of this genus usually occupy rich prairie soils. Most sunflowers tend to be rhizomatous and form colonies; this feature helps bind the soil. They are palatable to livestock and are rarely found in prairies which are overgrazed. Their nectar provides nourishment for nectarivores. Sunflower seeds provide food for many seed-eating small mammals and grassland birds. In the genus *Helianthus*, many species are allelopathic.

Helianthus mollis Lam., DOWNY SUNFLOWER

Height: 0.5–1 meter (1.5–3 feet)
Flowers: August–Mid September
Color: Yellow rays and disks
Habitat: Xeric

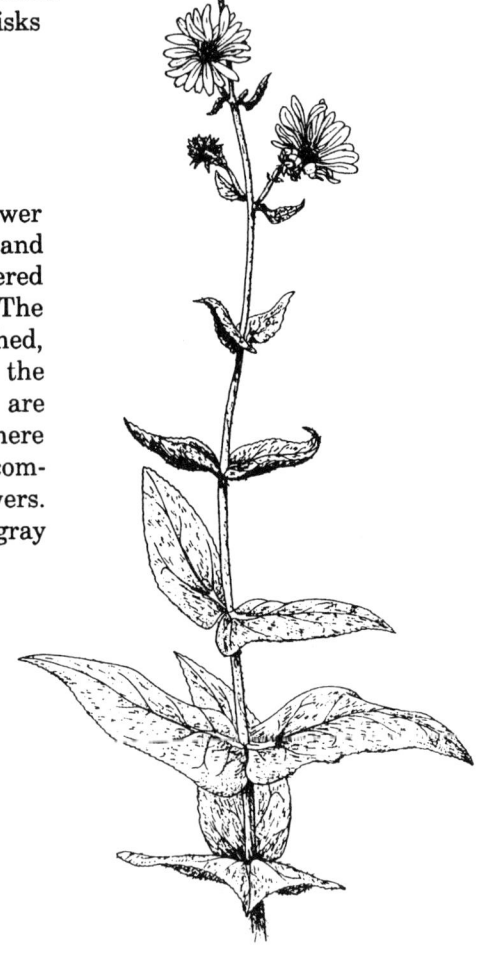

Identification Features:

The erect stems of Downy Sunflower arise from well-developed (strong and stout) rhizomes. They are covered with long, straight, soft hairs. The leaves are opposite, slightly toothed, rough to the touch, and clasp the stem. Both surfaces of the leaves are covered with soft white hairs. There are one to several flower heads composed of yellow ray and disk flowers. The entire plant has a whitish-gray appearance.

Helianthus occidentalis Riddell, WESTERN SUNFLOWER

Height: 0.5–1.5 meters (1.5–5 feet)
Flowers: Mid July–September
Color: Yellow rays and disks
Habitat: Xeric

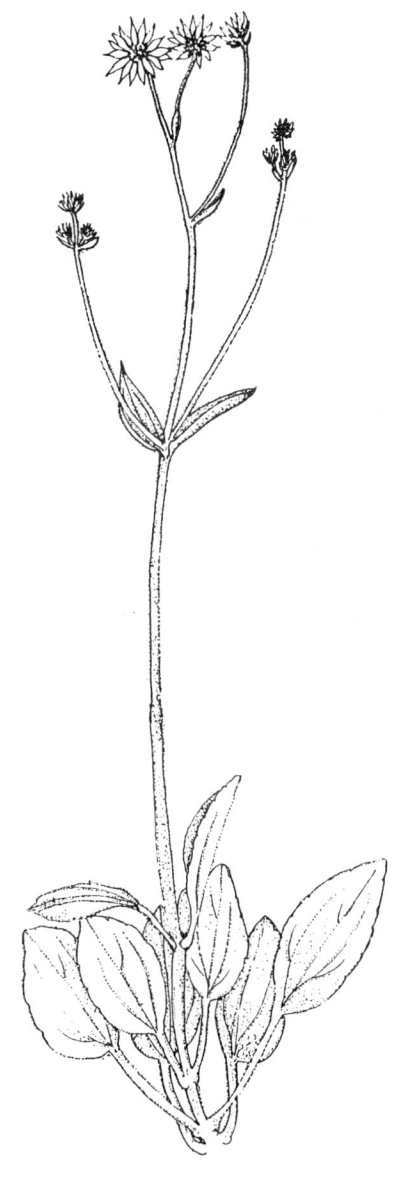

Identification Features:

The tall, erect, slender flexible stems of Western Sunflower arise from rhizomes. The leaves have three prominent veins and short hairs on both sides. The basal pairs of leaves are large and broadly ovate; the upper one or two pairs are small and lanceolate. The flower heads are 2–3 cm in diameter with ten to twenty yellow ray flowers and many yellow disk flowers.

The location and shape of leaves on the stem are major characteristics used to identify this sunflower. There are one or two pairs of small opposite leaves on the upper portion of the stem, usually none in the middle, and five or six pairs of large opposite leaves near the base.

Ecological Notes:

Western Sunflower is a characteristic species of sand and hill prairie. It is rhizomatous and forms vegetative colonies which may not flower. Its seeds are a favorite food item of goldfinches.

Helianthus rigidus (Cass.) Desf., PRAIRIE SUNFLOWER

Height: 0.5–1.5 meters (2–5 feet)
Flowers: August–September
Color: Yellow rays, Purplish-brown disks
Habitat: Mesic-Xeric

Identification Features:

Prairie Sunflower has rigid, erect, rough stems which arise from stout rhizomes. The stem is often tinged with a purple hue. The leaves are opposite, lanceolate, and tri-veined. They are stiff and rough. The flower heads contain fifteen to twenty yellow ray flowers and many purplish-brown disk flowers.

Some botanists classify this plant as *Helianthus laetiflorus rigidus* (Cass.) Fern. (F).

Heliopsis helianthoides (L.) Sweet, FALSE SUNFLOWER

Height:	0.6–1.5 meters (2–5 feet)
Flowers:	July–Mid September
Color:	Yellow-Orange rays and Yellow disks
Habitat:	Mesic-Hydric

Identification Features:

False Sunflower has an erect, smooth to scaly stem that arises from a short rhizome. Its leaves are opposite, ovate, prominently toothed, short-petioled, and rough to the touch on both surfaces. The flower heads are solitary to several and resemble a "sunflower." However, in true sunflowers only the disk flowers produce seeds. False Sunflower has fertile ray flowers, with forked pistils, in addition to the fertile disk flowers. The term *helianthoides* means "sunflower-like."

Ecological Notes:

This short-lived perennial species is common in prairie remnants and along thickets. False Sunflower has a tendency to form colonies.

Kuhnia eupatorioides L., FALSE BONESET

Height: 0.3–1 meter (1–3 feet)
Flowers: Mid August–Mid September
Color: Creamy-White disks
Habitat: Xeric

Identification Features:

The upright, sturdy and finely pubescent main stem of False Boneset grows from a stout taproot. The bush-like appearance of this species is due to the numerous upright branches of the main stem. The leaves are alternate, lanceolate, 2.5–10 cm long, and slightly pubescent. There are veins and gold-colored resin glands on the leaf's underside. The upper leaves are sessile, while the lower ones have short petioles. The leaves are alternate, not opposite as with the true bonesets of the genus *Eupatorium*.

The heads are in flat-topped clusters at the ends of the branches. Each head has seven to thirty tubular creamy-white disk flowers. There are no ray flowers. The feathery pappus of the achene is visually striking and provides wind dispersal.

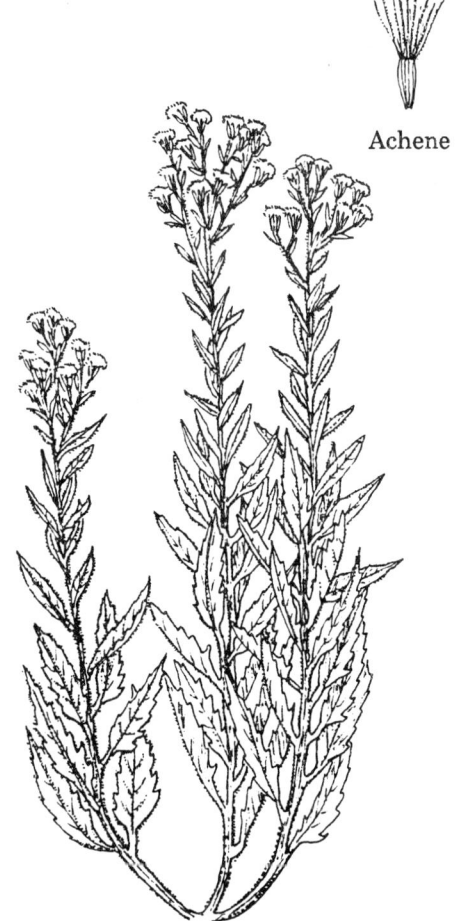

Achene

Ecological Notes:

False Boneset is palatable for ungulates during spring and early summer; however, it is avoided during the remainder of the growing season.

Ecological notes for the blazing stars, *Liatris* spp.

Blazing stars tend to be early colonizing plants that can thrive in degraded or recovering prairie communities. When compared to other early colonizers, they have a relatively long vegetative life. Blazing stars are well adapted to periods of drought. They have a deep root system for water absorption, small hairs that reflect sunlight, and numerous narrow leaves which help cool the plant. These anatomical features help reduce transpiration.

Inflorescences that occur on spikes for most plant species start to bloom at the base and progress upward; but blazing stars bloom from the top and proceed downward. While in bloom, the spike of rose-purple disk flowers is fuzzy with extended white stamens and pistils.

Most of the pollination observed for the blazing stars is carried out by bumblebees, a few butterfly species, beetles, flies, and carpenter bees. Bees rarely revisit flower heads while working an inflorescence, thus ensuring cross-pollination.

Various herbivores, in addition to insects, obtain energy from the blazing stars. Ungulates graze the young plants. Rodents, such as the meadow vole, seek out and eat the corms.

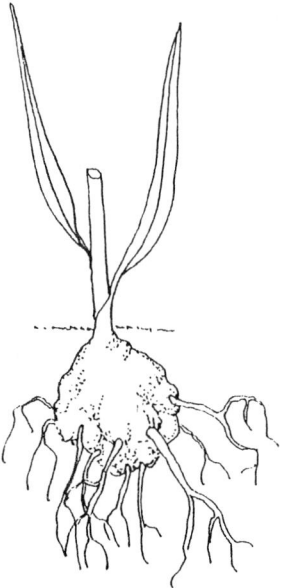

Blazing Star Corm

Liatris aspera Michx., ROUGH (BUTTON) BLAZING STAR

Height: 0.3–0.6 meter (1–2 feet)
Flowers: Mid August–September
Color: Rose-Purple disks
Habitat: Xeric-Mesic

Identification Features:

Rough Blazing Star has an erect stem that develops from a large corm. The alternate leaves are numerous, rough, and narrow along the entire length of the stem. They diminish in size toward the summit. The leaves and stems are pubescent. The involucral bracts are curved in at the edges and appear cup-shaped or "button-like." The flower heads of Rough Blazing Star are larger than those of Prairie Blazing Star but not as numerous.

involucral bracts

Head

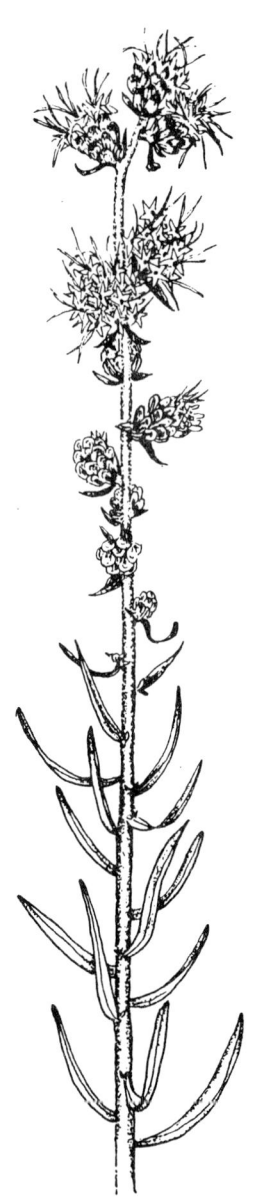

Liatris pycnostachya Michx., PRAIRIE BLAZING STAR

Height:	1–1.5 meters (3–5 feet)
Flowers:	Mid July–August
Color:	Rose-Purple disks
Habitat:	Mesic-Hydric

Identification Features:

The stem of Prairie Blazing Star is closely set with spirals of long, narrow leaves near the base that progressively get shorter toward the inflorescence spikelet. Its unbranched stem arises from a large corm that is 7–10 cm in diameter. Short, stiff hairs occur on both the leaves and stem. The involucral bracts are reddish, sharp-pointed, fuzzy, and bent out. The cylindrical inflorescence is a spike, 25–30 cm long, and crowded with sessile heads of disk flowers.

Liatris spicata (L.) Willd., MARSH BLAZING STAR, a closely related species, is distinguished from Prairie Blazing Star by its blunt-shaped, purple-tinged involucral bracts. This species grows best in hydric sites.

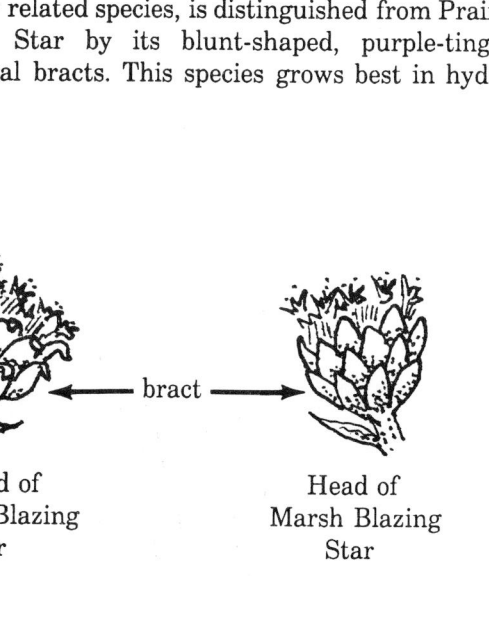

← bract →

Head of Prairie Blazing Star

Head of Marsh Blazing Star

Parthenium integrifolium L., WILD QUININE

Height: 0.3–1 meter (1–3 feet)
Flowers: Mid June–Mid August
Color: White rays and disks
Habitat: Mesic

Identification Features:

The stems of Wild Quinine arise from a thick tuberous taproot. Its leaves are rough, firm, serrated, and ovate-oblong in shape. The basal and lower leaves have long petioles and are large, up to 30 cm long and 10 cm wide. The upper ones are much smaller and clasp the stem. The beautiful white flower heads are numerous and in dense terminal clusters. Each flower head is about 4–5 mm in diameter with five tiny ray flowers and many disk flowers.

Ecological Notes:

Wild Quinine is a member of the stable prairie community and is rarely found in disturbed sites. Many small insects visit the white flower heads. The thick roots are a favorite food of voles.

Prenanthes aspera Michx., ROUGH WHITE LETTUCE

Height: 0.7–1.5 meters (2–5 feet)
Flowers: August–September
Color: Cream-Yellow rays
Habitat: Mesic-Xeric

Identification Features:

The stiff, upright stem of Rough White Lettuce grows from a swollen tuberous root. The entire plant is heavily pubescent. Large broad-lanceolate leaves are at the base; the upper leaves are reduced in size. The flower heads are crowded in small clusters at the upper part of the stem. The involucral bracts are pubescent and form a cylinder almost enclosing twelve to twenty ray flowers. There are no disk flowers.

Flower Head

Prenanthes racemosa Michx., GLAUCOUS WHITE LETTUCE, has smooth stems and leaves. It has pale-violet flowers. This species grows best in mesic to hydric soils.

Ecological Notes:

Prenanthes spp. are very important late season nectar sources for bees and other insects.

Ratibida pinnata (Vent.) Barnhart, YELLOW (GRAY-HEADED) CONEFLOWER

Height:	0.7–1.3 meters (2–4 feet)
Flowers:	July–August
Color:	Yellow rays, Greenish-Gray disks
Habitat:	Mesic-Xeric

Identification Features:

The stems of Yellow Coneflower are usually clustered, branched, and have deeply dissected leaves with three to seven segments. The stem arises from a shallow rhizome. The basal leaves have short petioles while the upper leaves are nearly sessile. All the leaves are smooth. The terminal inflorescence consists of four to ten drooping yellow ray flowers and a central head of greenish-gray disk flowers. Identification of this unique plant is easy due to its rather ragged appearance, gray-green deeply dissected basal leaves, drooping yellow ray flowers, and a central head of greenish-gray disk flowers.

Ecological Notes:

Yellow Coneflower is a short-lived, colonizing plant of the prairie community. It has a shallow rhizomatous root system that helps bind soil. Yellow Coneflower, when young, is highly palatable for ungulates.

Rudbeckia hirta L., BLACK-EYED SUSAN

Height: 0.3–0.6 meter (1–2 feet)
Flowers: Mid June–September
Color: Orange-Yellow rays, Brown disks
Habitat: Mesic

Identification Features:

Black-eyed Susan has highly branched, bristly-pubescent stems. Its alternate leaves are oblong, shallow-toothed, and pubescent. Black-eyed Susan has orange-yellow ray flowers and a central head of brown-black disk flowers studded with yellow pollen. Its inflorescences are up to 7 cm in diameter. Black-eyed Susan is very showy and remains in flower throughout the summer.

Ecological Notes:

Black-eyed Susan is a biennial or short-lived perennial that is an early colonizer of the prairie community. It often thrives in disturbed areas such as dusty roadsides. The leaves have long, stout hairs that serve to keep dust from clogging their breathing pores (stomata), thus helping to keep the plant alive and functioning properly.

Senecio pauperculus Michx., BALSAM RAGWORT

Height: 0.1–0.5 meter (4–20 inches)
Flowers: Mid May–June
Color: Yellow-Orange rays and disks
Habitat: Mesic-Hydric

Identification Features:

Balsam Ragwort is a small erect plant that arises from a branching rhizome. Its basal leaves are slender to oval with long petioles. The upper leaves are sessile and have deeply jagged and serrated lobes. There are few leaves on the stem. The flat-topped inflorescences at the tip of the stem are up to 2 cm in diameter. Both the ray and disk flowers are yellowish-orange. Balsam Ragwort adds much brightness to the late spring prairie scene.

Ecological Notes:

Balsam Ragwort can survive in disturbed areas and is also a member of the stable prairie.

Ecological notes for the *Silphium* species:

Silphiums are large, coarse forbs characteristic of the prairie community. The large root systems of silphiums are important in the soil building process as they can penetrate clay subsoil. The thick taproots, 2.5–5 cm in diameter, of Compass Plant and Prairie Dock descend vertically, start branching at 1 m down, and reach a depth of 3–4.5 m. The root systems of Rosinweed and Cup Plant are more fibrous and dense. As space between the soil and root increases due to root contraction during flowering, or during periods of drought, humus seeps into these crevices and black soil forms.

All silphiums are palatable and nutritious during early growth and are sought by ungulates. Silphiums look like sunflowers. They have both ray and disk flowers, all yellow. Silphiums produce large sunflower-like seeds that are actively sought by birds and rodents.

Moisture does not easily evaporate from the large silphium leaves due to their rough, waxy layer and orientation. In particular, the leaves of Compass Plant and Prairie Dock are erect above the ground and tend to present only their thin edges to the sun, thus reducing transpiration. The large leaves of these two plants feel cool even on hot days.

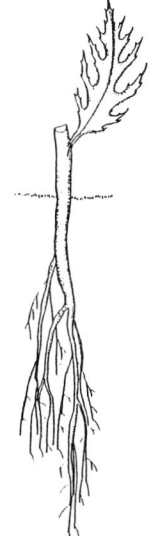

Compass Plant
Root

Compass Plant
Seeds
(actual size)

Silphium integrifolium Michx., ROSIN WEED

Height: 0.7–1.5 meters (2.5–5 feet)
Flowers: July–August
Color: Yellow rays and disks
Habitat: Xeric-Mesic

Identification Features:

The stem of Rosin Weed arises from rhizomes. Its leaves, as with other members of this genus, feel like sandpaper. They are opposite, lanceolate to ovate, and slightly toothed. The yellow terminal heads are approximately 5 cm in diameter. They are composed of both ray and disk flowers.

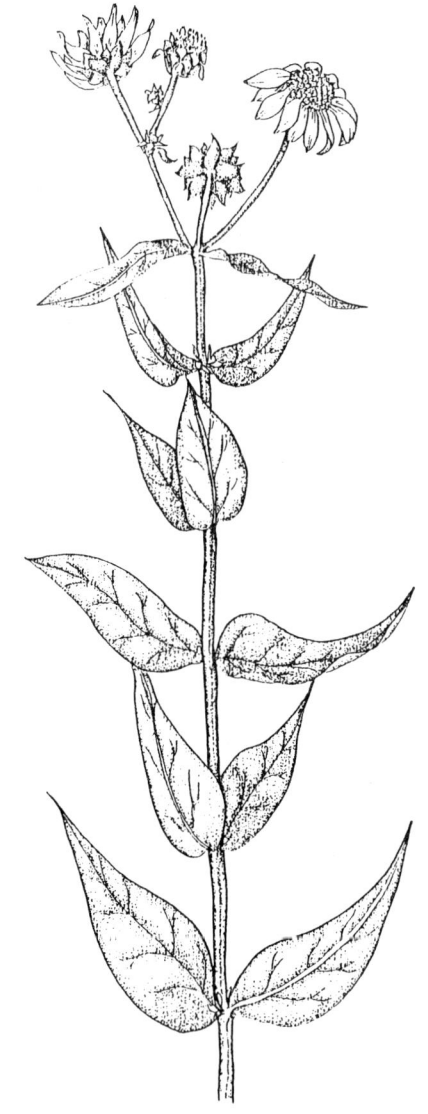

Silphium laciniatum L., COMPASS PLANT

Height: 1–2.5 meters (3–8 feet)
Flowers: July–August
Color: Yellow rays and disks
Habitat: Mesic-Xeric

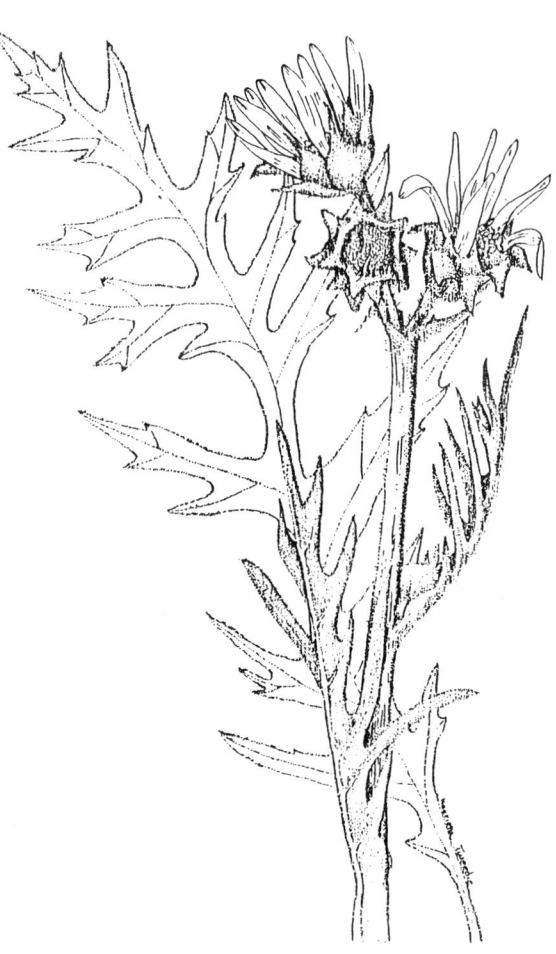

Identification Features:

The stem of Compass Plant is stout and bristly-hairy. The leaf edges, especially for plants that are not going to produce flowers, tend to point north and south, hence the common name. The basal leaves are large, deeply cut, and rough to the touch. The upper leaves of the stem are reduced. The plant is resinous throughout with globules of white resin often oozing out and sticking to the stem. The upper half of the stem contains large, 5–7 cm in diameter, bright yellow "sunflowers" with both ray and disk flowers.

Ecological Notes:

In the genus *Silphium,* Compass Plant has the greatest fidelity to undisturbed prairie communities. It is actively sought by native and domestic ungulates. A silphium weevil, *Rhynchites* sp., chews the stem near the flower heads so that they droop and wilt. Next it starts eating the disk flowers. Resin then oozes out of the stem, and the flower heads die.

Silphium perfoliatum L., CUP PLANT

Height: 1.2–2.1 meters (4–7 feet)
Flowers: Late July–August
Color: Yellow rays and disks
Habitat: Hydric

Identification Features:

The bases of the large opposite leaves are grown together and surround the square stem to form a "cup." The large toothed leaves are triangular-shaped. Cup Plant has numerous heads with both ray and disk flowers, all yellow.

Ecological Notes:

In addition to growing in prairies, Cup Plant also thrives in flood plains and degraded areas. Following a rain, considerable amounts of water are held in the "cups." Goldfinches and other birds have been observed drinking water from the "cups" of the Cup Plant.

Silphium terebinthinaceum Jacq., PRAIRIE DOCK

Height:	0.7–3 meters (2–10 feet)
Flowers:	July–September
Color:	Yellow rays and disks
Habitat:	Mesic-Hydric

Identification Features:

The leaves are 30 cm or longer, thick, resinous, coarsely toothed, spade-shaped, and rough like sandpaper. They are confined to the base of the plant and, like other silphiums, are coated with a rough, waxy layer. Usually during August, tall, smooth stalks arise from the cluster of leaves. At the top of these stalks there are about six smooth, round buds that open as small "sunflowers." The resin in the leaves and stems has an odor suggestive of turpentine, hence the Latin name *terebinthinaceum*.

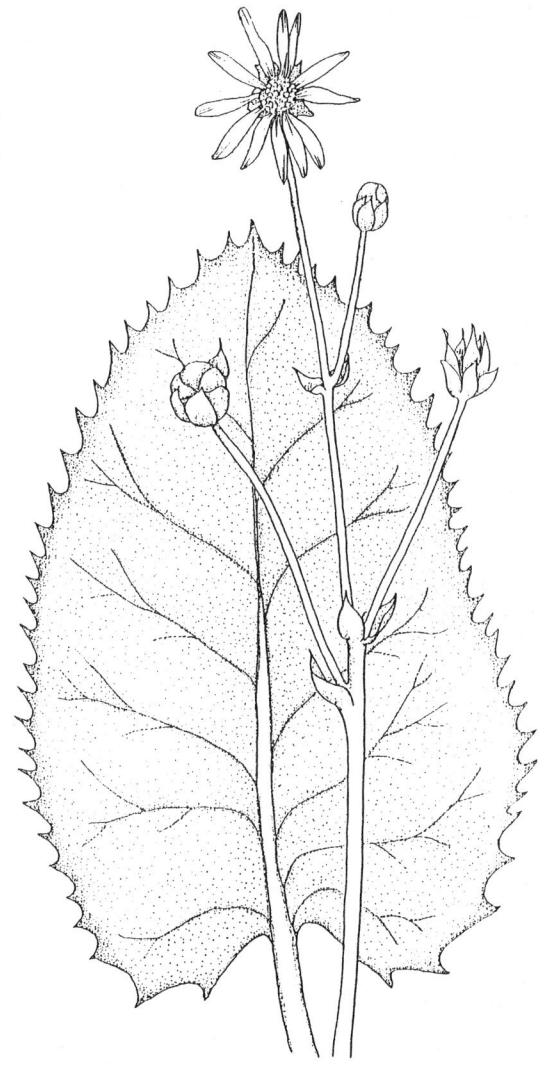

Ecological Notes:

A silphium weevil, *Rhynchites* sp., uses the buds of this plant as sites for egg laying and larval development. Goldfinches and other birds eat the seeds of Prairie Dock.

New shoots from the large taproot help Prairie Dock to survive severe degradation. It is sometimes the only remaining prairie plant species surviving in an area.

Ecological notes for the goldenrods, *Solidago* spp.

Goldenrods are common and hardy in prairie and in disturbed communities. The palatability of most goldenrods is low, and they often increase in overgrazed prairies. Goldenrods have a dense rhizomatous root system that helps bind the soil.

Certain species of soldier beetles, resembling brightly colored fireflies, feed on the goldenrod heads and carry their pollen from plant to plant. The pollen is sticky and spread by insects, not wind. It is *not* a source of hay fever as commonly accused.

Solidago missouriensis fasciculata
Holzinger, MISSOURI GOLDENROD

Height:	0.2–1 meter (0.6–3 feet)
Flowers:	August–Mid September
Color:	Yellow rays and disks
Habitat:	Xeric-Mesic

Identification Features:

The erect, smooth unbranched stem of Missouri Goldenrod arises from a strong creeping rhizomatous system. The stem is brown to green, sometimes even reddish. The leaves are alternate, narrow, sessile, stiff, and decrease in size as they ascend up the stem. They are pointed at the tip, serrated, and have three prominent parallel veins. Missouri Goldenrod's basal leaves are up to 15 cm long and are shed before flowering. The plume-shaped, recurved inflorescence supports a multitude of heads with tiny yellow ray and disk florets. The weight of the inflorescence tends to bend the stem.

Ecological Notes:

Colonies of this species produce several short, leafy stalks without flowers. It is a common goldenrod in dry prairies.

Solidago nemoralis Aiton, OLD-FIELD GOLDENROD

Height: 0.2–0.6 meters (0.5–2 feet)
Flowers: Mid August–September
Color: Yellow rays and disks
Habitat: Xeric

Identification Features:

Old-field Goldenrod has upright, unbranched, smooth stems that arise from rhizomes. The firm, narrow leaves are alternate, three-veined, slightly serrated, and weakly pointed at the tips. They are up to 15 cm long at the base and are much smaller toward the flower heads. The inflorescence is a recurving one-sided raceme with the heads pointing outward. The small heads have yellow ray and disk flowers.

Ecological Notes:

This species grows in old fields in addition to prairie. The dense rhizomatous root system of Old-field Goldenrod helps bind the soil. The narrow leaves of this species help radiate heat more readily, thus reducing some transpiration.

Solidago rigida L., RIGID (STIFF) GOLDENROD

Height: 0.3–1.7 meters (1–5 feet)
Flowers: Late August–September
Color: Yellow rays and disks
Habitat: Xeric-Mesic

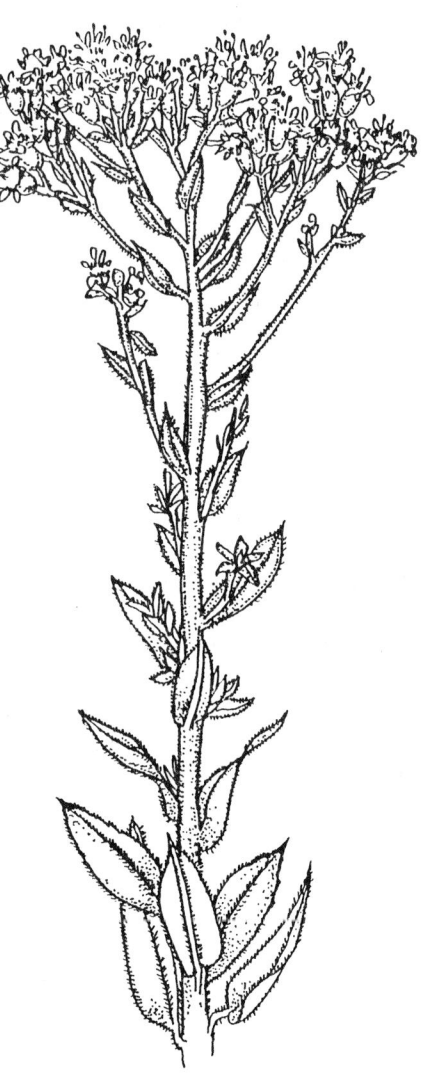

Identification Features:

Rigid Goldenrod has a rigid, hairy stem that arises from a short rhizome. Its stiff stem branches only at the top and is capped by dome-shaped to flat-topped clusters of small heads of yellow-orange ray and disk flowers. The inflorescence is about 15 cm in diameter. The alternate leaves are thick and leathery, rigid, pubescent on both surfaces, and have pointed tips. The large, lower leaves are up to 30 cm long and have short petioles, while the upper ones are smaller and clasp the stem.

Ecological Notes:

Rigid Goldenrod is a colonizing species that often invades disturbed areas. This species competes well with grasses for moisture as its fibrous roots can penetrate the soil to depths of 5 m. Rigid Goldenrod attracts many butterfly species when flowering.

Nitrogen fixation has been reported for Rigid Goldenrod but at a much lower rate than for legumes. It is unclear as to whether nitrogen fixation in this species is a biologically important source of nitrogen for individual plants.

Vernonia fasciculata Michx., COMMON IRONWEED

Height: 0.5–1.5 meters (2–5 feet)
Flowers: August–September
Color: Red-Violet disks
Habitat: Hydric-Mesic

Identification Features:

Common Ironweed has tough clustered stems which grow from rhizomes and fibrous roots. The smooth, reddish stems branch only at the inflorescence. The leaves are alternate, sessile, smooth on both sides, and numerous. They have sharply toothed margins and are conspicuously pitted on the undersides. The flower heads of red-violet disk florets form a flat-topped inflorescence. These brightly colored flowers add a splash of intense color to the prairie.

Ecological Notes:

Common Ironweed is an indicator of overgrazing; ungulates avoid this species because of its bitter taste and fibrous nature. *Vernonia* spp. are good nectar producers during the late summer months. Several species of this genus which are pasture weeds.

BORAGINACEAE (BORAGE)

Lithospermum canescens (Michx.) Lehm., PRAIRIE (HOARY) PUCCOON

Height:	0.2–0.3 meter (8–12 inches)
Flowers:	May–June
Color:	Yellow-Orange
Habitat:	Xeric-Mesic

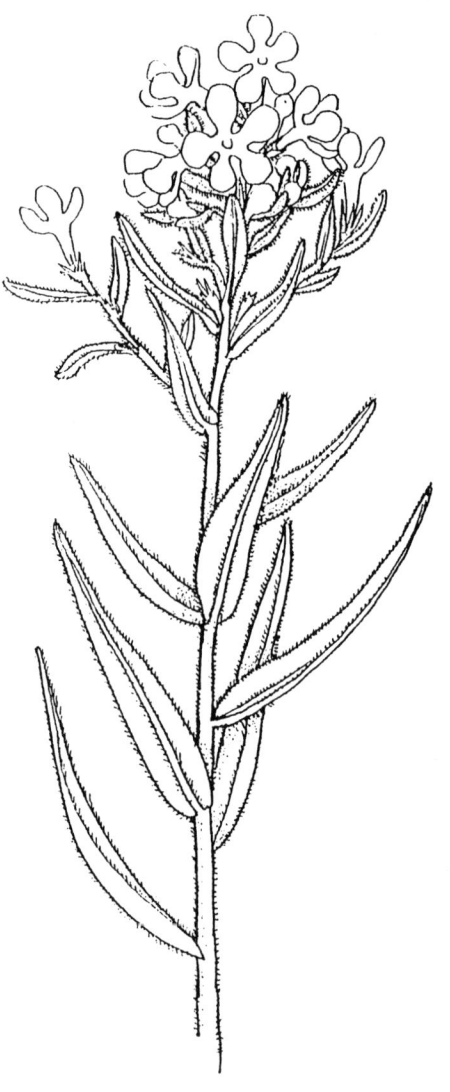

Identification Features:

One to several pubescent stems grow from a long taproot that contains reddish-purple juices. The alternate leaves are lance-shaped to oblong and are covered with short, stiff hairs. The stems are topped with inflorescences of yellow-orange, tubular flowers with five lobes. Prairie Puccoon glitters in sunlight! Its seeds are like little pieces of polished bone or ivory.

Ecological Notes:

The presence of Prairie Puccoon is an indication of virgin prairie conditions. The tube of the flower is constructed in such a manner that only a few species of insects, such as butterflies with long tongues, can enter to gather nectar.

Lithospermum croceum Fern., HAIRY PUCCOON, is found growing in sand prairie and Black Oak savanna.

CACTACEAE (CACTUS)

Opuntia humifusa Raf., PRICKLY PEAR

Height:	0.05–0.2 meter (2–8 inches)
Flowers:	Mid June–Mid July
Color:	Yellow
Habitat:	Xeric

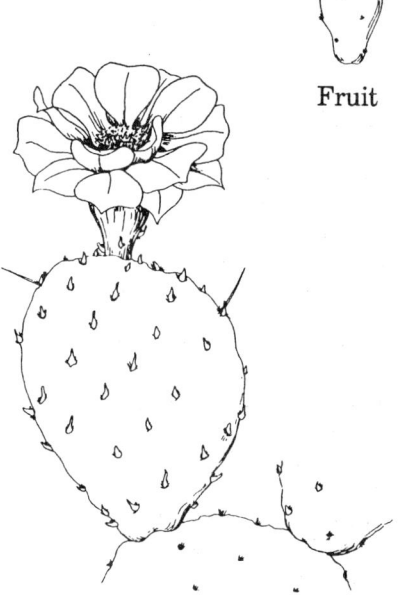

Fruit

Identification Features:

The prostrate, spreading stems or segments commonly known as "pads" of Prickly Pear arise from a branching fibrous root system. The pads are pale green, flat, thick, and succulent. They are often without spines; but may have one or two. Small, thick fleshy leaves are present for only a short period. The flowers have 8–12 waxy yellow petals and reddish centers. Stamens are numerous and very showy. The dull purple fruit is pulpy, egg-shaped, and contains many seeds.

Ecological Notes:

East of the Mississippi River, Prickly Pear is found growing mainly in sand prairie, Black Oak savanna, and rock outcrops. This cactus is normally restricted to small patches but like other *Opuntia* species, it can spread and form large mats in overgrazed pastures or during times of severe drought. The sharp spines on the pads may help protect some adjacent plants from being grazed by ungulates.

The water storing capabilities of the thick, succulent stems of *Opuntia* species make it extremely drought resistant. Most photosynthesis is carried out in the new pads. Reproduction occurs by both sexual and asexual means. Bees do much of the pollination. The seeds are dispersed by small mammals and birds. Roots form on the pads. Vegetative reproduction occurs when the pads break away from the parent plant and form new colonies.

COMMELINACEAE (SPIDERWORT)

Tradescantia ohiensis Raf., SPIDERWORT

Height: 0.3–1 meter (0.9–3 feet)
Flowers: Late May–July
Color: Bluish-Violet
Habitat: Mesic-Xeric-Hydric

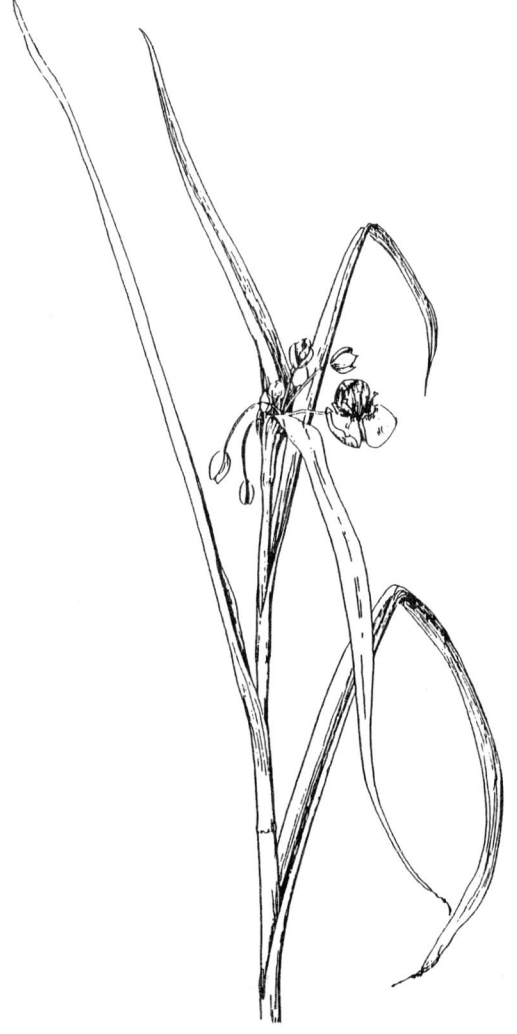

Identification Features:

The stems of Spiderwort arise from numerous fleshy roots. Its alternate leaves are narrow and long, up to 45 cm. They are folded lengthwise and curve downward. The stems contain a mucilaginous, stringy substance resembling that excreted by a spider. The flowers have three round to spade-shaped petals accented by six golden stamens.

Ecological Notes:

This plant needs sunshine, but not too much. During intense sunlight, the silky petals of the flower close. The mucilaginous, stringy substance enables Spiderwort to hold considerable amounts of water and may help it survive periods of drought. Ornate box turtles often use this plant species as a food and moisture source.

CYPERACEAE (SEDGE)

Carex bicknellii Britt., PRAIRIE (BICKNELL'S) SEDGE

Height:	0.3–0.8 meter (1–2.5 feet)
Flowers:	May
Color:	Yellowish anthers
Habitat:	Xeric-Mesic

Identification Features:

The erect stems of Prairie Sedge usually grow in narrow, upright tufts. Its stem has three angles. The stems of sedges are not round like those of grasses. The perigynium, an inflated sac that encloses the achene, is 4 mm or more in length.

Ecological Notes:

Prairie Sedge is a stable member of xeric and mesic prairie.

Perigynium

Carex meadii Dewey, MEAD'S SEDGE

Height:	0.2–0.5 meter (8–20 inches)
Flowers:	May
Color:	Greenish-Brown scales, Yellow anthers
Habitat:	Xeric-Mesic-Hydric

Identification Features:

The triangular stems of Mead's Sedge arise from elongated rhizomes. Most of the leaves are basal, smooth, and 3–7 mm wide. They are attached to the stems on all three sides. The terminal staminate spike is 1.5–3 cm long and is densely crowded with pollen producing flowers. Below this spike, there are one to three pistillate spikes which are 1–3 cm long. The egg-shaped perigynia have purplish-brown margins, prominent green midribs, and grow in six rows that are 3–4 mm long.

This species is barely distinguishable from *Carex tetanica* Schkuhr.

Ecological Notes:

Mead's Sedge grows in a variety of habitats, including hill prairie, fens, calcareous prairie, and calcareous swales.

Carex pensylvanica Lam., PENNSYLVANIA (OAK) SEDGE

Height: 0.2–0.4 meter (8–16 inches)
Flowers: April–Mid May
Color: Reddish-Brown scales, White-Yellow anthers
Habitat: Xeric-Mesic

Identification Features:

Pennsylvania Sedge grows in soft, low tufts. Its triangular stems arise from conspicuous stolons. The leaves are very slender, usually less than 3 mm wide. The staminate spike is reddish-brown, up to 2 cm long, and on a short stalk. There are one to four sessile egg-shaped pistillate spikes, 5–10 mm long. The perigynium is finely pubescent and is 2–3 mm long.

Ecological Notes:

This species grows in acidic soils of prairie and oak savanna. Pennsylvania Sedge provides important fuel for savanna fires.

Carex stricta Lam., STRICT (TUSSOCK) SEDGE

Height:	0.3–1 meter (1–3 feet)
Flowers:	May
Color:	Dark-Brown to Purple scales, White-Yellow anthers
Habitat:	Hydric

Identification Features:

Strict Sedge is found growing in large patches on elevated clumps. The lower portion of the stem is dark red to brown. The leaf blades are 2–5 mm wide. There are one to three staminate spikes which are 2–5 cm long. This *Carex* species is distinguished by its dark brown to purple staminate spikes above the overlapping greenish scales of the sessile pistillate spikes. There are two to six pistillate spikes. The common name comes from the strict or tight, narrow arrangement of staminate and pistillate spikes at the tops of stems.

Ecological Notes:

Strict sedge is common in sedge meadows and fens. The spotted turtle, *Clemmys guttata*, utilizes habitats dominated by cattails and this sedge species.

EUPHORBIACEAE (SPURGE)

Euphorbia corollata L., FLOWERING SPURGE

Height: 0.3–1.2 meters (1–4 feet)
Flowers: July–August
Color: White
Habitat: Xeric-Mesic

Identification Features:

The stem of Flowering Spurge is slender and smooth. Its leaves are oblong, 1–2 cm long, and scattered alternately along the stem. The base of each inflorescence has several leaves in a whorl. The flowers have petal-like appendages, but no true petals. The fruit is a broad capsule which breaks open with enough force to send the seed a considerable distance. The entire plant contains milky sap.

Ecological Notes:

Flowering Spurge can become slightly weedy in some disturbed prairie habitats. It is a poisonous plant which is seldom eaten by ungulates. Flowering Spurge has thick, milky sap which holds considerable amounts of water. The sap, along with its deep root system, helps this plant survive drought periods.

FABACEAE (LEGUME)

General importance of the legume family to the prairie and other notes:

The legumes are a major family in the prairie flora. Legume species aid in the soil building process by means of their deep penetrating root system and the nitrogen-fixing bacteria (Genus *Rhizobium*) which live in their root nodules. *Rhizobium* species convert inert nitrogen from the soil's atmosphere into compounds that are usable by the legumes and by adjacent members of other plant species. These nitrogen-fixing bacteria are genus-specific for the plant; for example, in the root nodules of lead plant, *Amorpha canescens,* live *Rhizobium amorpha.* The deep root systems of legumes aid in breaking up clay subsoil.

Some members of the legume family, such as the indigos, are very showy. Legumes produce hard seeds that remain viable in the soil for many years.

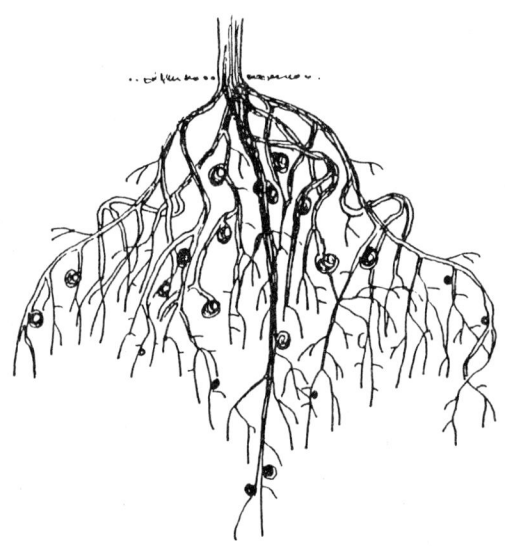

Legume Root System
with Nodules

Amorpha canescens Pursh, LEAD PLANT

Height: 0.6–1.3 meters (2–4 feet)
Flowers: Late June–July
Color: Blue-Violet-Purple
Habitat: Xeric-Mesic

Identification Features:

Lead Plant is a true prairie shrub. Its stems are woody and slightly pubescent. The alternate leaves have 15–51 small leaflets, 6–12 mm long. They are gray-green (lead) colored and are covered with fine hairs. The inflorescence is a narrow, crowded, highly branched raceme, 5–10 cm long. Its flowers show only a single petal and have ten bright yellowish-orange conspicuous anthers. Lead Plant provides striking beauty to the prairie throughout the growing season.

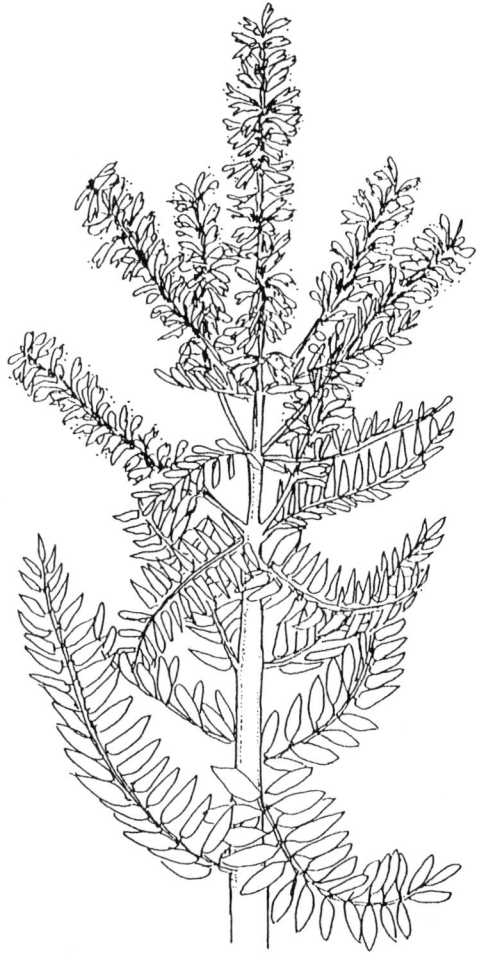

Ecological Notes:

Lead Plant is an indicator of virgin prairie. It is highly nutritious and palatable, and thus decreases under heavy grazing pressure. Lead plant moth, *Shinea lucens,* is one of many insects observed on this plant species. In the absence of fire or heavy grazing the stems become very woody.

Lead Plant has several anatomical features which help it survive periods of drought. These include: (1) a deep root system for water absorption, (2) lead-colored leaves which reduce the effect of solar heating by reflecting sunlight, and (3) finely divided leaflets which expose less surface area to sunlight, thus reducing transpiration.

Astragalus canadensis L., CANADIAN MILK VETCH

Height:	0.3–1.3 meters (1–4 feet)
Flowers:	July–August
Color:	Creamy to Greenish-White
Habitat:	Xeric

Identification Features:

The smooth stem of Canadian Milk Vetch is erect except for the lower portion which occasionally lies on the ground. It is stout and grows from rhizomes. The alternate leaves have thirteen to twenty-nine opposite leaflets. Cream-colored flowers are densely crowded in racemes arising from the upper leaf axils. The flower's keel (lowest petal of the pea flower) is short and rounded. The fruit is a plump and pointed pod with a beak. It is less than 1 cm in length and contains several tiny hard seeds.

Ecological Notes:

This *Astragalus* species is palatable to livestock. Certain other members of this genus which cause poisoning in livestock are called "locoweeds." The leaves of Canadian Milk Vetch tend to orient themselves in an east-west direction, thus preventing some evaporation.

Ecological notes for the indigos, *Baptisia* spp.

The indigos, *Baptisia* spp., are important soil builders. They have long penetrating roots which aid in breaking up clay subsoil. Living in their root nodules are the nitrogen-fixing bacteria, *Rhizobium baptisia,* which convert inert nitrogen into usable compounds. Although nutritious, the vegetative plant parts are rarely eaten due to their slight toxicity. Most of their large hard seeds are eaten or destroyed by insects. White Wild Indigo grows in prairie and savanna, whereas Cream Wild Indigo is found primarily in open prairie.

Cream Wild Indigo and White Wild Indigo are remarkable from both the standpoint of beauty and their insect-plant relationships. Described below are some pollination and seed predation features for these two prairie legumes.

Cream Wild Indigo blooms approximately two weeks earlier than White Wild Indigo. It is pollinated by bumblebee queens, *Bombus bimaculatus* and *Bombus nevadiensis auricomis,* that have just emerged from overwintering. The queens are also engaged in nest building, egg-laying, and rearing of immatures in addition to gathering nectar during this time. White Wild Indigo blooms approximately two weeks later and is generally pollinated by the worker bumblebee caste. Both species have a few visits from a honeybee, *Apis* sp. The wild-indigo duskywing butterfly, *Erynnis martialis,* lays its eggs on *Baptisia* spp. Ruby-throated hummingbirds have also been observed gathering nectar from White Wild Indigo, but they are not a major pollinator.

The nectar gathering pattern of bumblebee pollinators determines the likelihood of pollen transfer within the raceme. Indigos often have flowers in both the pistillate and staminate phase. Queens walk horizontally between the flowers of Cream Wild Indigo's raceme, moving from staminate phase flowers to pistillate phase flowers, and by so doing transfer pollen. Pollinator movement for White Wild Indigo is upward and reverse; the workers move from the pistillate to the staminate phase flowers.

White Wild Indigo produces a somewhat greater nectar reward than Cream Wild Indigo. Nectar production for White Wild Indigo is greater on days 3–4 of flowering than on day 1; there is no significant difference for Cream Wild Indigo. Frequency of pollination is usually higher in White Wild Indigo for three major reasons: (1) more bumblebees are present, (2) nectar rewards are higher, and (3) it is less subject to unpredictable weather conditions since flowering occurs later. As a result, more ovules (eggs) are fertilized and eventual seed predation is higher.

Several insects prey upon the seeds and pods of the indigos. Predation includes seed destruction, pod exit holes, webby material, and wall damage. In some years, there is almost a total loss of seeds except for a few isolated plants. The weevil, *Apion rostrum*, is the dominant seed predator. In early June, the weevils feed on the flower and leaves.

The female weevil drills holes into the base of the inflated pod throughout the flowering season, lays her eggs, and then pushes the eggs inside the pod with her snout. The egg is yellow and only slightly smaller than the seeds at this time. Individual weevils inhabit the pods for varying time periods. Exit holes, 1.5 mm in diameter, appear throughout the growing season; some adults leave when the pods dehisce (split), and others overwinter.

In some cases, the weevils may actually hasten seed germination by slightly gnawing the tough seed coat. Indigos disperse their seeds in a tumbleweed manner during late fall when the abscission layer at the stem base breaks. The stem, with its pods still attached, can then be blown by the wind.

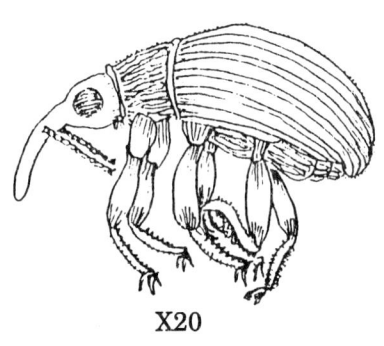

X20

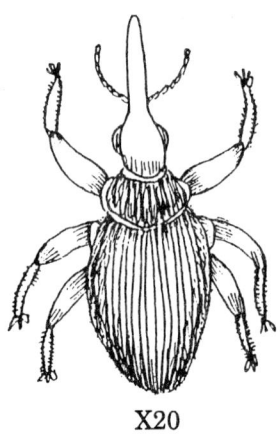

X20

Apion rostrum

Baptisia leucantha T. & G., WHITE WILD INDIGO

Height: 1–1.6 meters (3–5 feet)
Flowers: June–July
Color: White
Habitat: Mesic-Xeric

Identification Features:

White Wild Indigo has a smooth, stout stem with a widely branched crown. It resembles a miniature bushy tree. Its smooth leaves have small stipules, short petioles, and three leaflets, 1–2 cm long. The whole plant turns black upon drying. A tall, elegant inflorescence raceme, 30–60 cm, stands erect above the bushy plant. The inflorescence is closely set with white pea flowers about 2.5 cm long. The smooth, inflated legume pod is 2–3 cm long and has a short, tapering beak, up to 0.5 cm long.

Flower   Pod

Baptisia leucophaea Nutt., CREAM WILD INDIGO

Height:	0.5–0.9 meter (1.5–3 feet)
Flowers:	Mid May–Early June
Color:	Cream-Yellow
Habitat:	Mesic-Xeric

Identification Features:

This low, drooping bushy plant has two to twelve finely pubescent thick stems which are highly branched and widely spread, up to 0.8 m. The leaves are sessile, finely pubescent, and have two large stipules which give the appearance of leaflets. Cream Wild Indigo has only three leaflets which are up to 4 cm long, but there appear to be five due to the two large stipules. The inflorescence consists of a horizontal to drooping raceme of large cream-yellow pea flowers that are up to 3 cm long. The finely pubescent inflated pod is 4–5 cm long and has a long, tapering beak, up to 1.5 cm long. Cream Wild Indigo is a most handsome plant at all stages during the growing season.

Ecological notes for the tick trefoils, *Desmodium* spp.

The tick trefoils, *Desmodium* spp., have one of the most efficient seed dispersal mechanisms of any prairie species. The "hooked" hairs on the loment pod cling to any animal or person passing by and are thus dispersed. The seeds provide nourishment for grassland birds and small mammals. Nitrogen-fixing bacteria live in root nodules and convert inert nitrogen into usable compounds for the legume and adjacent plant species. The deep taproots aid in breaking up clay subsoil. Illinois Tick Trefoil has a greater fidelity to the prairie than Canada Tick Trefoil which can also be found in disturbed habitats.

Desmodium canadense (L.) DC., CANADA (SHOWY) TICK TREFOIL

Height:	0.6–1.8 meters (2–5 feet)
Flowers:	July–August
Color:	Purple
Habitat:	Mesic-Hydric

Identification Features:

Canada Tick Trefoil has a more bushy appearance than Illinois Tick Trefoil. The stems and leaflets are highly pubescent and have hooked hairs. Its leaf petioles are shorter than those of Illinois Tick Trefoil. The inflorescences are highly branched racemes of purple flowers on short pedicels. The flowers are numerous and dense. The legume pod is a loment that breaks into one-seeded sections with hooked hairs.

Loment

Desmodium illinoense Gray, ILLINOIS TICK TREFOIL

Height: 0.6–1.7 meters (2–5.5 feet)
Flowers: July–Mid August
Color: Pink-Light Purple
Habitat: Xeric-Mesic

Identification Features:

The tall erect stem of Illinois Tick Trefoil is finely pubescent and has hooked hairs. Its alternate leaves have a long petiole and three leaflets. They are pubescent with hooked hairs beneath. The inflorescence is a tall, sparsely branched raceme with pink to light purple pea flowers. The legume pod is a loment that breaks into one-seeded sections with hooked hairs.

Loment

Lespedeza capitata Michx., ROUND-HEADED BUSH CLOVER

Height: 0.6–1.5 meters (2–5 feet)
Flowers: Mid August–Mid September
Color: White
Habitat: Mesic-Xeric

Identification Features:

Round-headed Bush Clover has an erect, silvery-pubescent stem. Its alternate leaves have three leaflets which are oblong to lance-like in shape. They are relatively smooth on the upper side but silvery-pubescent beneath. The inflorescences are clustered heads, up to 3 cm in diameter, of small cream to white pea-shaped flowers near the top of the stems. The seed heads turn brown in the fall.

Ecological Notes:

Round-headed Bush Clover grows in prairie and savanna. It is excellent forage for ungulates and decreases with heavy grazing. The seeds provide food for grassland birds.

Ecological notes for the prairie clovers, *Petalostemum* spp.

Petalostemum species are indicators of virgin prairie. They are highly sought after by herbivores, such as rabbits and ungulates, and decrease with grazing pressure. The prairie clovers have anatomical features that help them survive in times of water stress. Their finely divided leaflets offer less surface area to the sun; thus the plant is cooler and transpiration is reduced. A wide-ranging root system from the taproot helps absorb water. Nitrogen-fixing bacteria, *Rhizobium petalostemum*, in the root nodules convert inert nitrogen into usable compounds for the clovers and neighboring members of other plant species.

Petalostemum candidum (Willd.) Michx., WHITE PRAIRIE CLOVER

Height:	0.3–0.6 meter (1–2 feet)
Flowers:	Late June–July
Color:	White
Habitat:	Xeric-Mesic

Identification Features:

White Prairie Clover has few to several smooth, upright stems arising from a taproot with widespread branching roots. The alternate leaves have five to nine linear-oblong, dotted leaflets. The leaflets are broader (more than 2 mm wide) than those of Purple Prairie Clover. The terminal inflorescence consists of one to a few spikes of crowded, small, white flowers.

Some botanists classify this plant as *Dalea candida* Willd.

Petalostemum purpureum (Vent.) Rydb., PURPLE PRAIRIE CLOVER

Height: 0.3–0.6 meter (1–2 feet)
Flowers: July
Color: Purple
Habitat: Mesic-Xeric

Identification Features:

The slender, erect, wiry stems of Purple Prairie Clover arise from a short vertical taproot that is highly branched. The alternate leaves have three to five leaflets which are much narrower (less than 2 mm wide) and closer together than those of White Prairie Clover. The inflorescence is a firm cylindric spike of crowded, small, purple flowers with five protruding bright orange-yellow anthers. Purple Prairie Clover is a desirable plant wherever it grows.

Some botanists classify this plant as *Dalea purpurea* Vent.

Psoralea tenuiflora Pursh, SCURFY PEA

Height:	0.5–1 meter (1.5–3 feet)
Flowers:	Mid June–July
Color:	Lavender
Habitat:	Xeric

Identification Features:

Scurfy Pea has a highly branched, grayish-pubescent, slender stem that arises from a stout taproot. Its alternate leaves are on long petioles and have three to five linear-oblong leaflets with glandular dots. The inflorescences of small, lavender flowers (3 mm) are on racemes up to 6 cm long.

Ecological Notes:

Scurfy Pea, like Lead Plant, can survive extreme drought conditions because of anatomical features such as small, gray-colored leaflets which reduce transpiration. Many insects infest this legume including a weevil, *Apion* sp., which destroys the flower bud. An abscission layer at the stem base breaks during late summer allowing the stem and seeds of Scurfy Pea to disperse in a tumbleweed manner.

Vicia americana Willd., AMERICAN VETCH

Height: 0.3–1.0 meter (1–3 feet)
Flowers: Mid May–June
Color: Bluish-Purple to Rose
Habitat: Hydric-Mesic

Identification Features:

The slender, smooth, weak stems of American Vetch often climb over other plants by means of their tendrils. Each leaf has eight to eighteen opposite leaflets and a tendril at the tip. There are sharp, serrated stipules at the base of each leaf. Racemes of three to nine flowers arise from the leaf axils. The pea flowers have a short upper lip consisting of three petals and a longer lower lip consisting of two petals. The pods are 2.5–3.5 cm long. Upon drying, they split lengthwise and drop their seeds.

Ecological Notes:

American Vetch is excellent forage for herbivores such as cattle, deer, and pheasants. It is a member of stable prairie and savanna communities.

Tendril

GENTIANACEAE (GENTIAN)

Gentiana andrewsii Griseb., BOTTLE (CLOSED) GENTIAN

Height:	0.2–0.6 meter (0.6–2 feet)
Flowers:	September–October
Color:	Dark-Light Blue
Habitat:	Hydric-Mesic

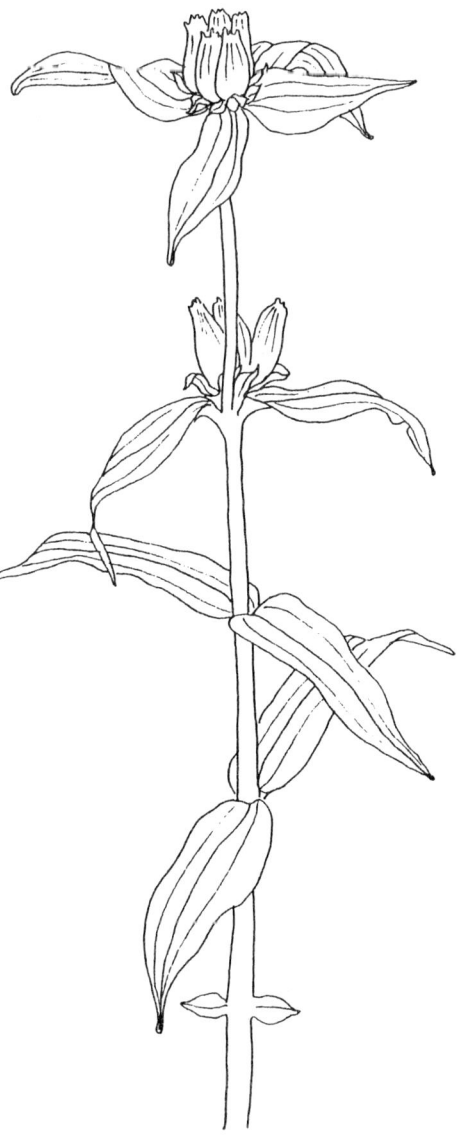

Identification Features:

Bottle Gentian has a relatively smooth stem that arises from a short rootstock. There are long internodes between the opposite pairs of smooth, parallel-veined leaves. The lance-shaped leaves do not have petioles and tend to bend downward. Tufts of flowers are clustered in the upper leaf axils. The united corolla is nearly closed or "bottle-shaped." The seeds are small, white-cream colored, and winged.

Ecological Notes:

The bumblebee is one of the few insects strong enough to open the bottle-shaped flower and achieve pollination. Attracted by the blue color, the bumblebee presses on the tip of the united corolla and pushes its front half into the "bottle." The entrance behind the bumblebee is held open with its abdomen and rear legs so that the bee does not become trapped.

Gentiana flavida Gray, YELLOWISH (CREAM) GENTIAN

Height: 0.3–0.6 meter (1–2 feet)
Flowers: Mid August–September
Color: Yellowish-Cream
Habitat: Xeric-Mesic

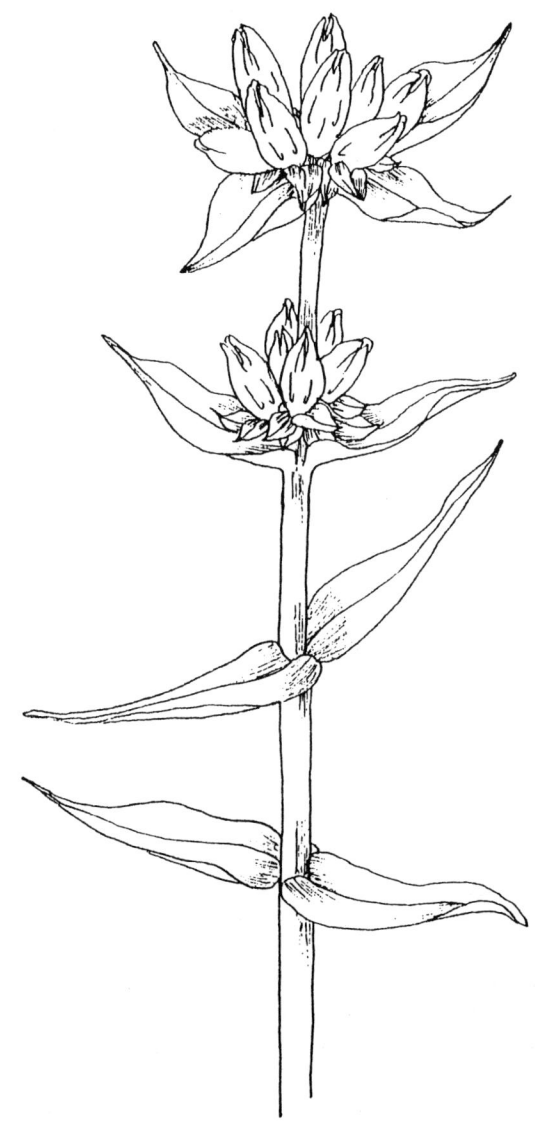

Identification Features:

The stem of Yellowish Gentian is smooth. The internodes between the pairs of leaves are shorter than those of Bottle Gentian; thus the foliage is more clustered than that of Bottle Gentian. Tufts of yellowish-cream flowers are clustered in the upper leaf axils. The fruits are papery capsules that contain small, whitish, winged seeds.

Ecological Notes:

Yellowish Gentian thrives quite well in both undisturbed and disturbed prairie. It is also a member of the savanna community.

Gentiana puberulenta Pringle, DOWNY (PRAIRIE) GENTIAN

Height:	0.2–0.5 meter (8–20 inches)
Flowers:	Mid September–October
Color:	Blue-Purple
Habitat:	Xeric-Mesic

Identification Features:

Downy Gentian has one to few slender, stiff, minutely pubescent stems which grow from a taproot with lateral roots. There are thirteen to nineteen opposite pairs of leaves along the stem. They are stiff, 2–7 cm long, sessile, and lanceolate to oblong-ovate. The leaves are longer as they ascend up the stem; those near the inflorescence tend to be in whorls of four to six.

Blue-purple flowers are crowded at the summit in the upper axils. The calyx tube is about 1 cm long. The corolla is 3.5–5 cm long and funnel-shaped; it has five prominent lobes that are pointed, folded, and reflexed. The fruits are papery capsules that contain small, whitish, winged seeds. Downy Gentian is one of the most beautiful wildflowers of the prairie.

Ecological Notes:

The presence of Downy Gentian indicates virgin prairie as this species tolerates little disturbance. This plant prefers calcareous soils. The flower's stigma does not fully develop until the large yellow anthers have shriveled up, thus ensuring cross-fertilization. The flowers open only on sunny days.

IRIDACEAE (IRIS)

Iris virginica shrevei (Small) E. Anderson, BLUE FLAG IRIS

Height:	0.3–0.9 meter (1–3 ft)
Flowers:	June–July
Color:	Lavender-Blue-Violet
Habitat:	Hydric

Identification Features:

The stems of Blue Flag Iris arise from extensive colonies of horizontal rhizomes with fibrous roots. Two or three parallel veined leaves, 0.4–0.5 m long and 2–3 cm wide, clasp the stem at the base. At the top of the stem, there are one or two beautiful flowers above the leaf-like bracts. There are three down curving petal-like sepals. Each sepal is blue-violet and has a bold yellow midrib with a hairy, bright yellow splotch near its base. The petals are blue-violet and erect; they are narrower than the sepals and lack any yellow color. Three petal-like styles arch over the stamens. This arching of the styles prevents self-fertilization. The fruit is a thick, three-lobed capsule that is 3–4 cm long.

Ecological Notes:

Blue Flag Iris is a common plant of calcareous fens, marshes, and wet prairie. The yellow color of the sepals attracts insects.

Sisyrinchium albidum Raf., BLUE-EYED GRASS

Height:	0.1–0.25 meter (4–10 inches)
Flowers:	Mid May–June
Color:	Blue-White
Habitat:	Xeric-Mesic

Identification Features:

The slender, wiry flattened stems and "grass-like" foliage of Blue-eyed Grass arise from a rootstock. When mature, this species grows in tufts. The common name refers to the flowers which seem to be unnaturally attached to the stems near the tips. Although normally blue, Blue-eyed Grass can be white-flowered in alkaline soil.

Ecological Notes:

Blue-eyed Grass thrives best in stable prairies but can survive in disturbed prairie. The underground rootstocks may be eaten by herbivores that "root" for food.

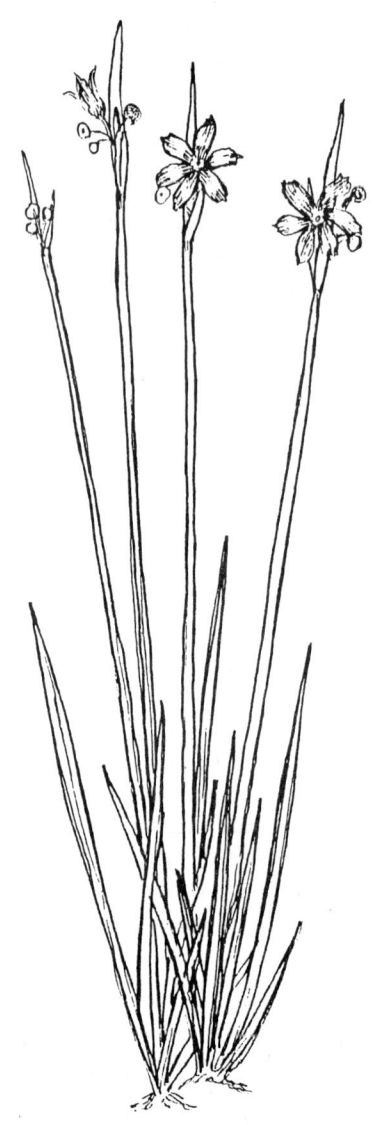

LAMIACEAE (MINT)

Monarda fistulosa L., WILD BERGAMOT (BEEBALM)

Height:	0.6–1.2 meters (2–4 feet)
Flowers:	July–August
Color:	Lavender
Habitat:	Hydric-Mesic-Xeric

Identification Features:

The stem of Wild Bergamot is square, slightly hairy, and arises from a branched rhizome. Its leaves are opposite, ovate-lanceolate shaped, and sharply serrated along their margins. The inflorescences are solitary, terminal heads that have many clusters of two-lipped tubular flowers. Both the leaves and flowers are dotted with glands that secrete volatile, aromatic bergamot oils.

Ecological Notes:

Wild Bergamot often grows in large colonies in prairie, disturbed habitats, and along forest edges. It is pollinated by insects, primarily bumblebees, honeybees, and wasps. The clear-wing sphinx moth also visits this plant. Pollination in Wild Bergamot is interesting due to two major observations. First, inflorescences are composed of several clusters, each with ten or more flowers open at a time, with some flowers in the staminate and others in the pistillate phase. However, young stigmas have a delayed receptivity to accepting self-pollen. Second, bumblebees and honeybees appear to visit staminate and pistillate phase flowers indiscriminately for nectar. Successful cross-pollination and outbreeding of Wild Bergamot is due, at least in part, to the continuous opening of the flowers during the day and the stigma's receptivity to cross-pollen prior to self-pollen.

Physostegia virginiana (L.) Benth., FALSE DRAGONHEAD (OBEDIENT PLANT)

Height: 0.3–1.5 meters (1–5 feet)
Flowers: August–September
Color: Pink-Rose
Habitat: Mesic-Hydric

Identification Features:

False Dragonhead has a square, smooth, stout stem. Its leaves are opposite, lanceolate to oblong, and sharply toothed. The lower leaves are 5–12 cm long and 5–20 mm wide; the upper leaves are smaller. The flowers are 2–3 cm long and are borne in terminal leafless spikes up to 20 cm long. Their funnel-shaped corolla resembles a "dragon's head." Each corolla has a two-lobed upper lip and a spotted three-lobed lower lip; the middle lobe of the lower lip is larger, rounded, and notched.

When a flower is pushed to one side it remains there, hence the alternate name, "obedient plant."

Ecological Notes:

False Dragonhead colonies are formed by rhizomes. Pollination is by insects.

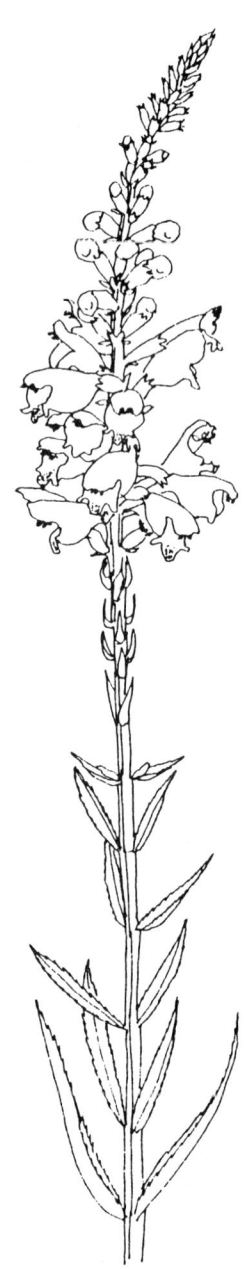

Pycnanthemum virginianum (L.) Duran & Jackson, COMMON MOUNTAIN MINT

Height:	0.6–0.9 meter (2–3 feet)
Flowers:	Mid July–August
Color:	White
Habitat:	Hydric-Mesic

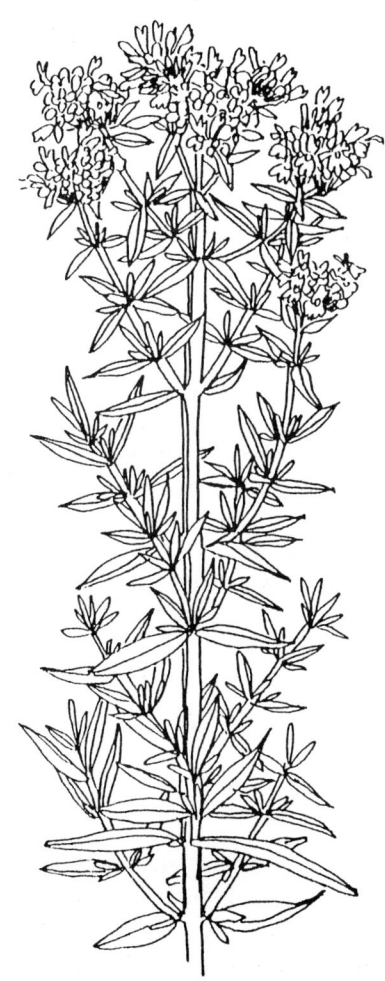

Identification Features:

The stem of Common Mountain Mint is four-angled or "square." Its narrow leaves are opposite, numerous, and lance-shaped. The white-purplish flowers are in head-like clusters about 5 mm in diameter. Common Mountain Mint is the most aromatic plant of the prairie. When crushed, its plant parts have a strong mint-like odor.

Ecological Notes:

In addition to inhabitating hydric-mesic and sometimes xeric prairies, Common Mountain Mint also occurs in habitats with considerable degradation. Its rhizomes extend a few inches from the parent plant and produce new shoots. These new shoots form tight colonies which appear as clumps. Many species of small insects are attracted to this plant.

LILIACEAE (LILY)

Allium cernuum Roth, NODDING WILD ONION

Height:	0.3–0.5 meter (1–1.5 feet)
Flowers:	Mid July–August
Color:	White-Pink-Lavender
Habitat:	Mesic

Identification Features:

The stems and leaves of Nodding Wild Onion arise from underground bulbs which are smaller than domestic onions. The leaves are usually long, narrow, tubular, and may be either hollow or solid. White, pink, or lavender flowers are borne at the tip of leafless stalks in solitary rounded clusters that "nod." When young, the flowering stalk bends in the middle. This bend or "nod" grows toward the top of the stem as the plant matures. The seeds of Nodding Wild Onion are black and hard. When crushed, the plant parts have the distinctive onion odor.

Ecological Notes:

Nodding Wild Onion prefers calcareous soils. It can form colonies by vegetative reproduction. Its succulent herbage is grazed by ungulates and its bulbs are dug up by herbivores that "root" for food.

Camassia scilloides (Raf.) Cory, WILD HYACINTH

Height:	0.2–0.4 meter (8–16 inches)
Flowers:	May–Early June
Color:	White to Pale-Purple
Habitat:	Mesic-Hydric

Identification Features:

The leafless flower stem of Wild Hyacinth grows from a black-coated bulb that is 2–4 cm long. The basal grass-like leaves have a prominent midrib on the underside and grow up to 30 cm. There are ten to twelve star-shaped flowers along the upper end of the raceme. Each white to pale purple flower has three petals and three sepals of the same color. There are six prominent yellow stamens in each flower.

Ecological Notes:

Wild Hyacinth is a member of prairie and savanna communities. It often forms dense colonies.

Hypoxis hirsuta (L.) Coville, YELLOW STAR GRASS

Height:	0.05–0.1 meter (2–4 inches)
Flowers:	Mid May–Mid June
Color:	Yellow
Habitat:	Mesic-Hydric

Identification Features:

The slender floral stem of Yellow Star Grass grows from a tuft of basal, "grass-like" leaves which arise from a small, hairy corm. One to several flowers arise from each slender stalk. The flower has three yellow petals and three yellow sepals. It blooms for about a day. Yellow Star Grass is a delicate, charming plant.

Ecological Notes:

Yellow Star Grass grows in woods, moist calcareous meadows, and mesic-hydric prairies.

Lilium michiganense Farw., TURK'S CAP LILY

Height: 0.6–1.5 meters (2–5 feet)
Flowers: Late June–July
Color: Orange-Red
Habitat: Hydric-Mesic

Identification Features:

Turk's Cap Lily has a stout, erect stem that grows from a yellowish, scaly, fleshy bulb. The bulb has a tendency to produce shoots from which a new bulb may develop; thus this species is often found growing in colonies. Smooth, lance-shaped leaves are 5–15 cm long, tapered at both ends, and grow in whorls of four to twelve around the stem. The large flowers face downward. Both sepals and petals are strongly recurved. They are yellow at the base and have several purple-brownish spots. A single flower may bloom up to four weeks.

The exotic *Lilium tigrinum* L., TIGER LILY, occasionally escapes cultivation and can be mistaken for Turk's Cap Lily. Tiger Lily has small dark bulblets in the leaf axils. The presence of these bulblets distinguishes Tiger Lily from Turk's Cap Lily.

Ecological Notes:

Turk's Cap Lily is found growing in prairie, in savanna, along streams, and at the forest's edge.

Lilium philadelphicum andinum (Nutt.) Ker, PRAIRIE LILY

Height: 0.3–1 meter (1–3 feet)
Flowers: June–Early July
Color: Reddish-Orange
Habitat: Xeric-Mesic

Identification Features:

The erect stem of Prairie Lily grows from a white, scaly, fleshy bulb that is 2–3 cm in diameter. The leaves near the top of the stem are sessile, 5–10 cm long, and are arranged in whorls of three to seven; they are alternate toward the base. Each plant produces one to five very showy flowers which are erect and open toward the sky. Purple spots decorate the inside base of each petal. The anthers are dark purple and very attractive. This beautiful species is very rare.

Ecological Notes:

This true prairie species cannot tolerate disturbance.

LINACEAE (FLAX)

Linum sulcatum Riddell, GROOVED YELLOW FLAX

Height: 0.3–0.7 meter (1–2 feet)
Flowers: August–September
Color: Yellow
Habitat: Xeric

Identification Features:

Grooved Yellow Flax has a pale green stem that is erect, slender, wiry, and conspicuously grooved. Its pale green leaves are alternate, sessile, narrow, and less than 2.5 cm long. The leaves usually shed early, leaving two tiny stipule glands on the stem where they were attached. Saucer-shaped flowers are borne on short stalks from the upper leaf axils. They have five yellow blunt petals and two rows of sepals at their base. Each plant bears many flowers, but only a few open each day. They open in the early morning but wither and fall off as the sun becomes hot later during the same day. The seeds are in small, hard yellow capsules which split open when mature.

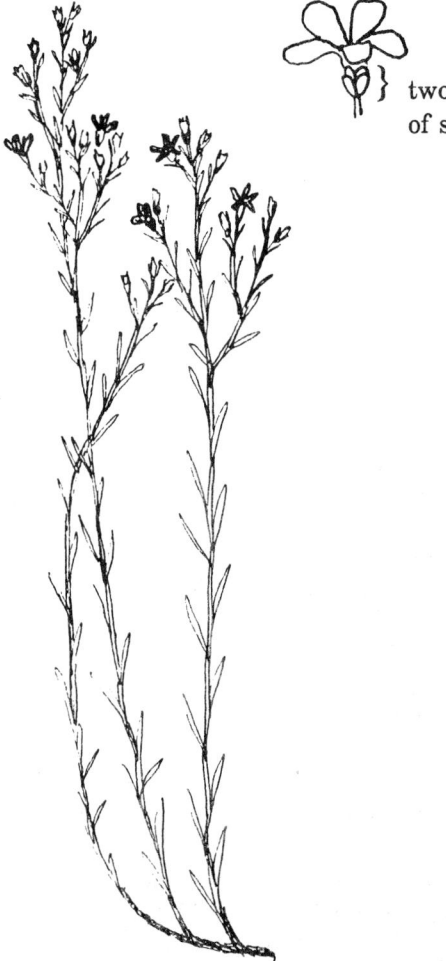
} two rows of sepals

Ecological Notes:

Nearly all of the prairie plant species are perennial. Grooved Yellow Flax is one of the few annuals of stable prairie communities. This species grows in dry hill and gravelly prairies where there is sparse vegetation; space is available for its seeds to germinate.

LOBELIACEAE (LOBELIA)

Lobelia spicata Lam., PALE SPIKED LOBELIA

Height:	0.2–0.6 meter (8–24 inches)
Flowers:	June–July
Color:	Pale-Blue to White
Habitat:	Mesic-Hydric

Identification Features:

Pale Spiked Lobelia has a slender, erect stem that grows from a fibrous root system. The stems are smooth near the top, but densely pubescent near the base. The leaves are alternate, oblong, shallowly toothed, and nearly sessile. They are 5–10 cm long at the stem's base but are reduced to bracts toward the summit. There are small tubular flowers, 9–12 mm in length, in the bracts of the raceme. Each corolla is two-lipped; the upper lip has two erect lobes while the lower lip is deeply cleft into three lobes. Five small green sepals surround the base of each flower.

Ecological Notes:

Although sensitive to grazing, Pale Spiked Lobelia sometimes survives in areas invaded by Eurasian plants. Animals tend to avoid grazing this species because it contains several alkaloids which taste bitter.

Flower

ONAGRACEAE (EVENING PRIMROSE)

Oenothera pilosella Raf., PRAIRIE SUNDROPS

Height:	0.2–0.6 meter (8–24 inches)
Flowers:	June–July
Color:	Yellow
Habitat:	Hydric-Mesic

Identification Features:

The slender, stout stem of Prairie Sundrops has hairs which are 2–3 mm long. The leaves are alternate, sessile, and lanceolate to oblong. Bright yellow flowers are located at the tips of stems and branches. The calyx tube is 0.5–2.5 cm long. The petals are 1–3 cm long. Eight bright yellow stamens with anthers, 4–8 mm long, are present. The pistil has a distinctly noticeable four-lobed or cross-shaped stigma. The four-angled fruit capsule splits and releases seeds over a period of time.

Ecological Notes:

Prairie Sundrops is a member of the stable prairie community.

ORCHIDACEAE (ORCHID)

Cypripedium candidum Willd., WHITE LADY'S SLIPPER

Height:	0.15–0.3 meter (6–12 inches)
Flowers:	May
Color:	White
Habitat:	Hydric

Identification Features:

The stiff, green stalks of White Lady's Slipper are clasped by finely pubescent leaves that have parallel veins. Its leaves ascend up the stem. The small, waxy, white "slipper" has green-purple sepals and petals extending outward and over it. The "slipper" has purple veins. Inside the white pouch there are purple speckles. A beauty to behold!

Ecological Notes:

White Lady's Slipper thrives best in moist, alkaline, black-soil prairies. There are symbiotic relationships between its roots and soil fungi. Most pollination is accomplished by tiny bees, 6–7 mm long, that are attracted to the fragrant, showy lip of this flower. Once inside the flower, it may take the small bee, or an occasional wasp or beetle, up to 15 minutes to crawl out of this beautiful blossom. While exiting, the bee gets smeared with sticky, green pollen and carries it to the next blossom. Since the blossom lacks nectar, the insect receives no nourishment for its efforts.

Habenaria leucophaea (Nutt.) Gray,
PRAIRIE WHITE FRINGED ORCHID

Height:	0.3–0.9 meter (1–3 ft)
Flowers:	Late June–July
Color:	Creamy-White
Habitat:	Hydric

Identification Features:

The stout stem of Prairie White Fringed Orchid grows from a partially fleshy tuberous rootstock with fibrous roots. The lanceolate leaves are 10–20 cm long and 2–3 cm wide. They are smooth, alternate, parallel-veined, and clasp the stem at the base. The lower leaves are considerably larger than the upper ones. Up to forty flowers are borne on a raceme that is about 7 cm long. Each flower is 2–3 cm long and has three sepals and three petals. The three sepals are all alike, and along with two lateral petals form a hood. The lower petal forms a lip which is deeply cut into three feathery, wedge-shaped lobes. Each lobe is divided into deeply fringed filaments which give a featherly appearance to the flower. A nectar spur, 3–5 cm long, extends back from the lower petal.

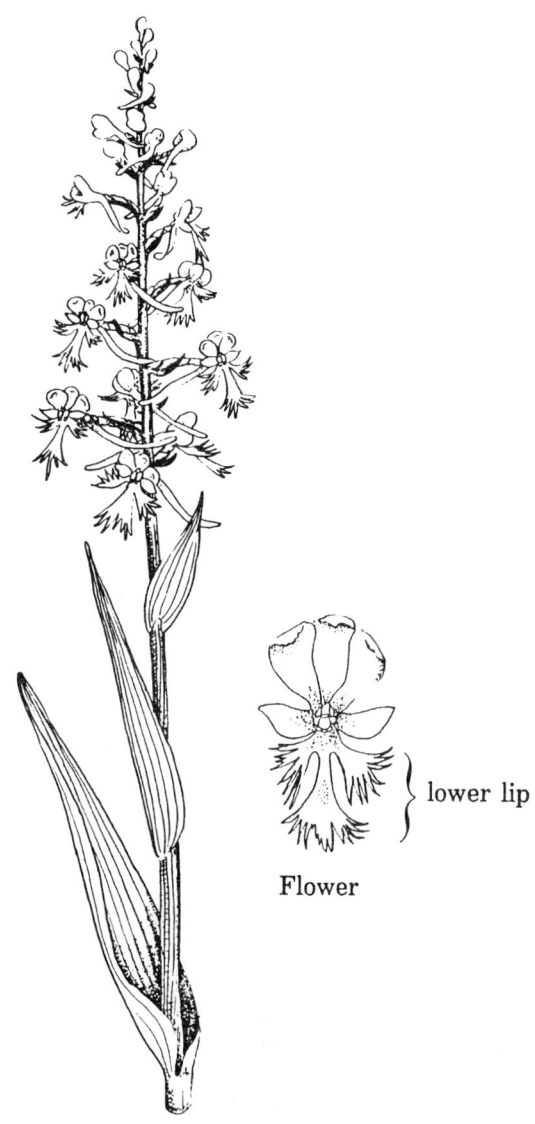

Flower

Ecological Notes:

Prairie White Fringed Orchid is an extremely rare species that requires wet calcareous soils. It has nearly been exterminated by cultivation and drainage. Dead, accumulated vegetation greatly suppresses the flowering of this species; Prairie White Fringed Orchid is rarely noticed unless the prairie has been recently burned.

Pollination is accomplished by wasps and butterflies during the day and by a variety of night flying visitors, including sphinx or hawk moths. The insects are attracted by the nectar spur's scent, and can land on the orchid's landing platform. While the insects are seeking nectar, pear-shaped clumps of pollen become stuck on their heads and are subsequently transferred to other plants.

OXALIDACEAE (OXALIS)

Oxalis violacea L., VIOLET WOOD SORREL

Height:	0.1–0.2 meter (4–8 inches)
Flowers:	May–Early June
Color:	Rose-Violet
Habitat:	Mesic-Xeric

Identification Features:

Violet Wood Sorrel does not have a true stem. Its leaves and flower stalks arise directly from a brown, scaly bulb that is about 1 cm in diameter. The leaves are 5–13 cm long and have three heart-shaped or "clover-like" leaflets. Each leaflet folds along a center crease and has a notch at the tip of its outer margin. The leaflets are smooth, bluish-green above and reddish-purple below. The flower stalks are up to 20 cm tall and contain several flowers with five rose-violet petals.

Ecological Notes:

In addition to prairie, Violet Wood Sorrel thrives in savanna. Its leaflets fold downward at dusk or in cloudy weather. All parts of the plant contain oxalic acid which tastes sour and may give this species some protection from insect grazers. When mature, the capsules rapidly invert and explosively scatter their seeds.

POACEAE (GRASS)

General importance of the grass family to the prairie and other notes:

The grasses contribute most of the prairie plant biomass and provide fuel for prairie fires. Humus from the decay of their deep and extensive root systems is mainly responsible for the rich, black color of grassland soils.

The growth habit of prairie grasses is sod-forming or bunch-forming, or both. Sod-forming grasses, such as cordgrass and switch grass, reproduce sexually from seed and asexually from underground stems called rhizomes. The rhizomes extend horizontally from a few inches to a few feet from the parent plant and produce new shoots from their tips or from the nodes along the stem. This results in a dense stand of shoots that may completely occupy the soil. A few sod-forming grasses also reproduce from horizontal stems above the ground called stolons.

Bunch-forming grasses have an erect growth of many shoots at their base. These shoots are called tillers. Little bluestem and prairie dropseed are bunch-forming grasses. Like sod-forming grasses, bunch-forming grasses reproduce sexually and asexually. Sometimes a grass species, such as big bluestem, may have both sod-forming and bunch-forming growth habits.

Besides the taxonomic classification, prairie grasses may be grouped according to their climatic origin as either "cool-season" or "warm-season" grasses. Cool-season grasses, such as needlegrass and bluejoint, are of northern origin. They renew their growth in early spring and usually mature from April to early July. Cool-season grasses become semi-dormant during the hot summer months and renew growth during autumn. By contrast, warm-season grasses, such as big and little bluestem, begin growth during late spring and grow continuously during the summer and into early fall. They mature during mid-fall. Warm-season and cool-season grasses intermingle freely as their requirements overlap.

Grasses have several anatomical features which enable them to flourish in the relatively harsh environment of the prairie. Some of these anatomical features include:

(1) deep root systems that reach depths of 2–4 m and branch profusely, enabling the grasses to obtain water during periods of drought,

(2) tough stems, reinforced by a silicon oxide, which enable the grasses to withstand high winds, and

(3) specialized "hinge cells" in the upper epidermis of the leaf.

During periods of drought they lose water rapidly, then contract and cause the leaf to roll up in a long tube. The pores through which water vapor is normally transpired are now inside the tube. The exposed lower surface of the leaf is highly resistant to water loss.

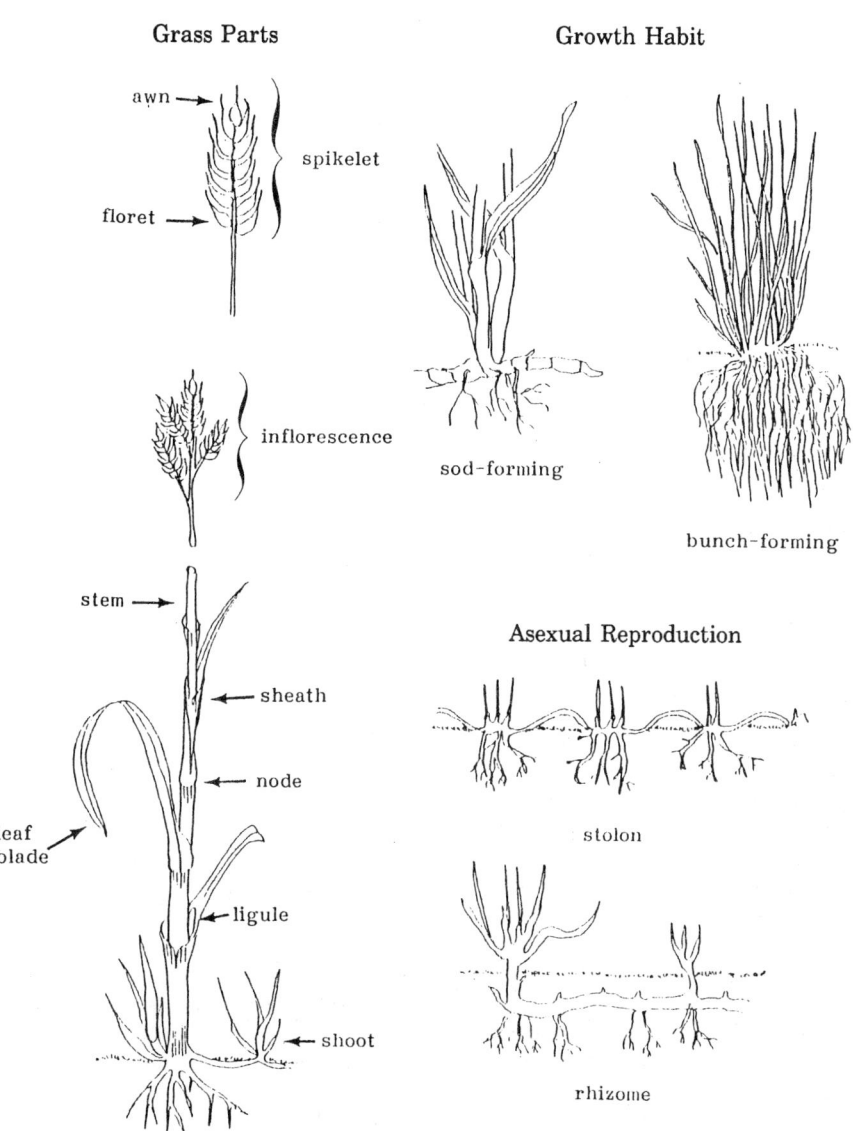

Andropogon gerardii Vitman, BIG BLUESTEM (TURKEY FOOT)

Height: 1–3 meters (3.5–10 feet)
Flowers: August–Mid September
Color: Yellow anthers
Habitat: Mesic-Xeric-Hydric

Identification Features:

Big Bluestem has both sod-forming and bunch-forming growth habits. The leaves of Big Bluestem remain green for most of the summer. During late summer and early fall its wax-like stem coating turns bluish-purple, and following a frost the foliage turns reddish-purple. Big Bluestem takes its popular common name from the bluish-purple color of the stem and nodes; however, it is also called "turkey foot" because its seed spikes frequently branch into three parts. Its numerous leaves may be up to 1.2 cm wide and are often hairy. The yellow anthers of its flowers are showy.

Ecological Notes:

This warm-season grass is one of the most widely spread and important dominants of the tallgrass prairie. Reasons for its dominance include rapid growth, dense sod-forming habit, height, and shade tolerance of mature plants and seedlings. In lowland tallgrass communities, Big Bluestem often grows in nearly pure stands.

Butterfly larvae of dusted skipper, *Atrytonopsis hianna;* ottoe skipper, *Hesperia ottoe;* and beard-grass skipper, *Atrytone arogos,* feed on Big Bluestem. Big Bluestem has a high protein content and is highly palatable for both native and introduced ungulates. Its deep fibrous root system reaches depths of 1.5 to 2.2 m and branches profusely. Humus from decay of its root system is very important in soil formation, and greatly contributes to the rich, black color of grassland soil. Big Bluestem provides much fuel for prairie fires.

Andropogon scoparius Michx., LITTLE BLUESTEM

Height: 0.5–1.1 meters (1.5–3.2 feet)
Flowers: Late August–September
Color: Pinkish-White seed spikes
Habitat: Xeric-Mesic

Identification Features:

Little Bluestem is a strong bunch-grass. The basal shoots and stem nodes are bluish-purple. The leaves are more slender and wiry than those of Big Bluestem. Other features distinguishing Little Bluestem from Big Bluestem are a somewhat flattened basal portion of stems and leaf sheaths, its slightly folded leaves, and its lack of pubescence on both the sheaths and lower portions of leaves. During fall, the pinkish colored stems have fuzzy and fluffy, white-silvery seed spikes.

Many botanists classify this plant as *Schizachyrium scoparium* Nash.

Ecological Notes:

This warm-season grass grows in many habitats with soils ranging from deep to shallow and rocky, and sandy to clay textured. It is an upland dominant but intermingles with tall grasses on lower sites. The roots reach depths of 1.4–1.8 m, are very abundant, and form a dense sod. Little Bluestem is a keen competitor as it possesses a deep and extensive root system along with a reduced transpiring leaf blade surface. These characteristics enable Little Bluestem to survive periods of drought.

Butterfly larvae of ottoe skipper, *Hesperia ottoe,* and crossline skipper, *Polites origenes,* feed on Little Bluestem.

Bouteloua curtipendula (Michx.) Torr., SIDE-OATS GRAMA

Height: 0.4–1 meter (1.3–3 feet)
Flowers: Late July–August
Color: Crimson-Red anthers
Habitat: Xeric

Identification Features:

Side-oats Grama takes its name from the oat-like florets which appear to hang from one side of the seed stalk. It retains the zigzag-topped seed stalks into autumn. The leaves are normally flat and have stiff hairs along the bases of their edges. While maturing, the tips of the upper leaves die as the basal leaves curl and turn light colored. The flowers have beautiful, showy, crimson-red anthers.

Ecological Notes:

Side-oats Grama is a warm-season, sod-forming grass which prefers xeric upland communities with alkaline conditions. Its dense and well-branched root system, extending to 0.5 m deep, can occupy deep or shallow soil. Side-oats Grama does not tolerate shade from the taller grasses. It is a weak grass species competitor unless the area is heavily grazed, gravely, or extremely dry. Under these conditions, its short rhizomes promote the formation of small areas of sod.

Bromus kalmii Gray, PRAIRIE BROME

Height:	0.3–0.7 meter (1–2 feet)
Flowers:	Mid June–Mid July
Color:	Inconspicuous
Habitat:	Mesic

Identification Features:

Prairie Brome is a solitary grass with soft, drooping inflorescences that are somewhat open and spreading. The leaf blades are flat and the edges of the sheath grow together to form a tube.

Ecological Notes:

This cool-season grass thrives best in calcareous soils. Prairie Brome is a member of prairie and savanna communities.

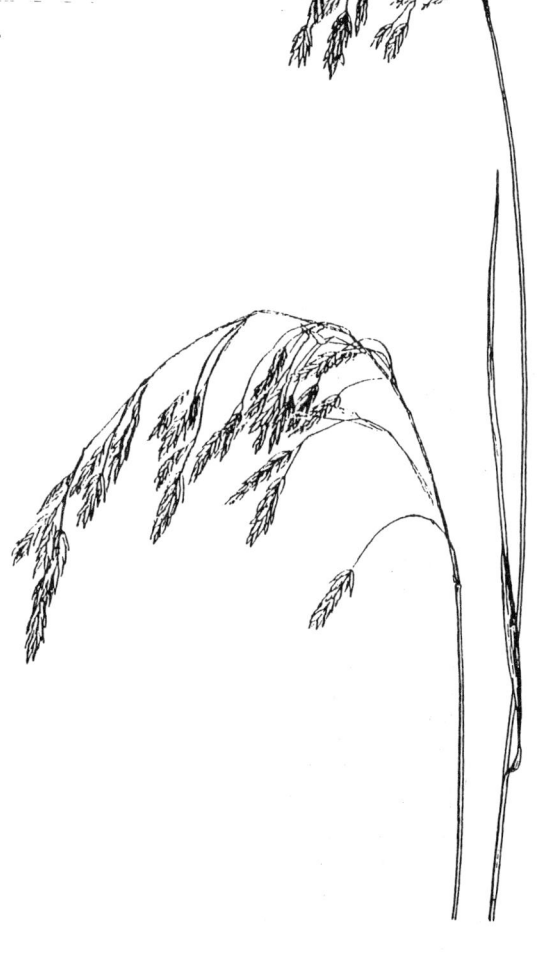

Calamagrostis canadensis (Michx.) P. Beauv., BLUE JOINT GRASS

Height: 0.6–1.2 meters (2–4 feet)
Flowers: June–Mid July
Color: Inconspicuous
Habitat: Hydric

Identification Features:

Blue Joint Grass has long, beautiful blue-green leaf blades and grows in dense clumps. The inflorescences are approximately 15 cm long and have white, silky hair on one side. The nodes between leaf blades are usually swollen.

Ecological Notes:

This cool-season grass often forms solid stands, eliminating other species, especially in marshy sites. Blue Joint Grass is nutritious and relished by ungulates.

Elymus canadensis L., CANADA WILD RYE

Height: 1–2 meters (3–7 feet)
Flowers: July–August
Color: Amber
Habitat: Xeric-Mesic

Identification Features:

During early summer, Canada Wild Rye is recognized by the green, waxy color of its pointed leaves. As the plant matures, it develops nodding seed spikes, up to 15 cm, with long, recurved awns. Its leaves clasp the stems by means of auricles (ear-like lobes). The leaf blades are 2.5 cm wide and up to 30 cm long. The upper blade surface is rough to the touch and tends to curl inward toward the tip. As Canada Wild Rye matures, its seed spikes and leaves turn russet to tan.

Ecological Notes:

This cool-season, short-lived grass grows in clumps in a wide variety of habitats. Canada Wild Rye is an early colonizing plant; its shallow root system provides protective cover in disturbed areas. It is highly palatable for ungulates but is not tolerant of heavy grazing. The seed spikes persist throughout the winter and provide nourishment for wildlife.

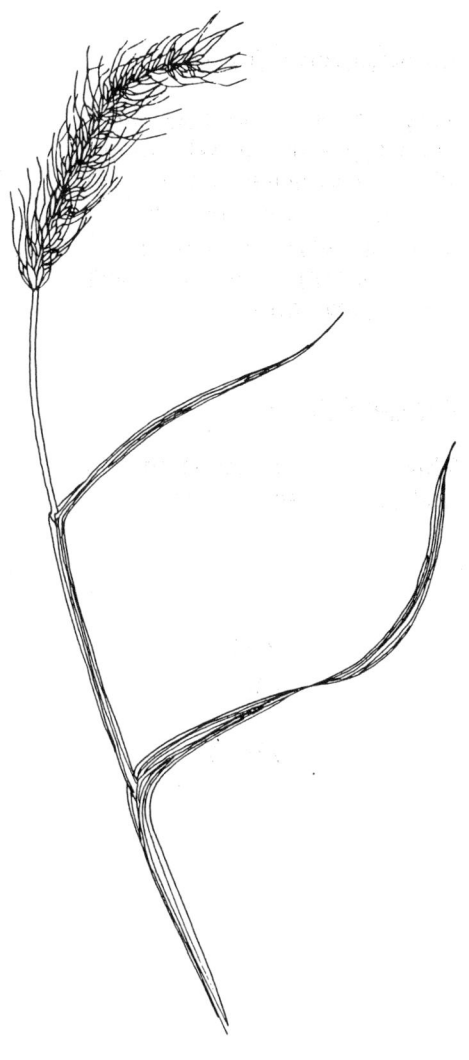

Hierochloe odorata (L.) Beauv., VANILLA GRASS

Height: 0.3–0.6 meter (1–2 feet)
Flowers: May
Color: Amber spikelets
Habitat: Hydric

Identification Features:

This fragrant, vanilla-scented grass forms dense sod clumps. The inflorescences are 7–8 cm long and to one side. The leaf blades are small while the plant is in bloom; however, long blades grow later.

Ecological Notes:

This cool-season grass thrives best in calcareous, moist soils.

Spikelet

Koeleria cristata (L.) Pers., JUNE GRASS

Height: 0.3–0.6 meter (1–2 feet)
Flowers: June–Mid July
Color: Silvery-Green
Habitat: Xeric

Identification Features:

June Grass grows in tufts, 5–7 cm in diameter. Its narrow stems and leaves arise from a fibrous root system. The leaves are stiff, up to 30 cm long, less than 3 mm wide, and rough on the upper surface. When maturing, they become curled and twisted. The dense spike-shaped, symmetrical inflorescence is 7–13 cm long. After the seeds are mature, the entire plant becomes golden yellow.

Ecological Notes:

June Grass occurs in scattered clumps in sand prairie, Black Oak savanna, and open-shrub communities. This short-lived, cool-season grass is an associate of Little Bluestem. It provides excellent forage but decreases with heavy grazing.

Panicum oligosanthes scribnerianum (Nash) Fern.,
SCRIBNER'S PANIC GRASS

Height:	0.2–0.3 meter (8–12 inches)
Flowers:	June
Color:	Reddish-Purple stigmas
Habitat:	Xeric

Identification Features:

Scribner's Panic Grass has a low lying branched appearance and grows in scattered tufts. The pointed leaf blades are approximately 0.8 cm wide and 6 cm long. Fine hairs are present on the undersides of the leaves. The spikelets are short, covered with fine hairs, and produce brilliant reddish-purple stigmas.

Ecological Notes:

Scribner's Panic Grass is a cool-season grass of Black Oak savanna and sand prairie. It cannot tolerate grazing.

Panicum leibergii (Vasey) Scribn., PRAIRIE PANIC GRASS, a closely related species, is found only in the finest mesic black soil prairie remnants.

Spikelet

Panicum virgatum L., SWITCH GRASS

Height:	1–2 meters (3–6.2 feet)
Flowers:	August–Mid September
Color:	Reddish-Purple anthers
Habitat:	Mesic-Xeric-Hydric

Identification Features:

Switch Grass is a sod-forming grass. It has strong glossy leaves. Identification is simplified by the somewhat inverted V-shaped wedge of fine, wavy hairs along the leaf's upper surface where its leaf blade joins the sheath. The large, open panicles are up to 40 cm long and begin to unfold in July. Its small inconspicuous flowers have reddish-purple anthers. Large tear-drop shaped seeds are borne on the open panicles.

Ecological Notes:

Switch Grass is a warm-season species that thrives in a wide range of habitats. Its roots are 3–4 mm in diameter and penetrate almost vertically to depths of 2 to 3.7 m. Switch Grass is aggressive, often forming dense stands and excluding other species.

Butterfly larvae of Leonard's skipper, *Heperia leonardus,* and tawny-edge skipper, *Polites themistocles,* eat Switch Grass.

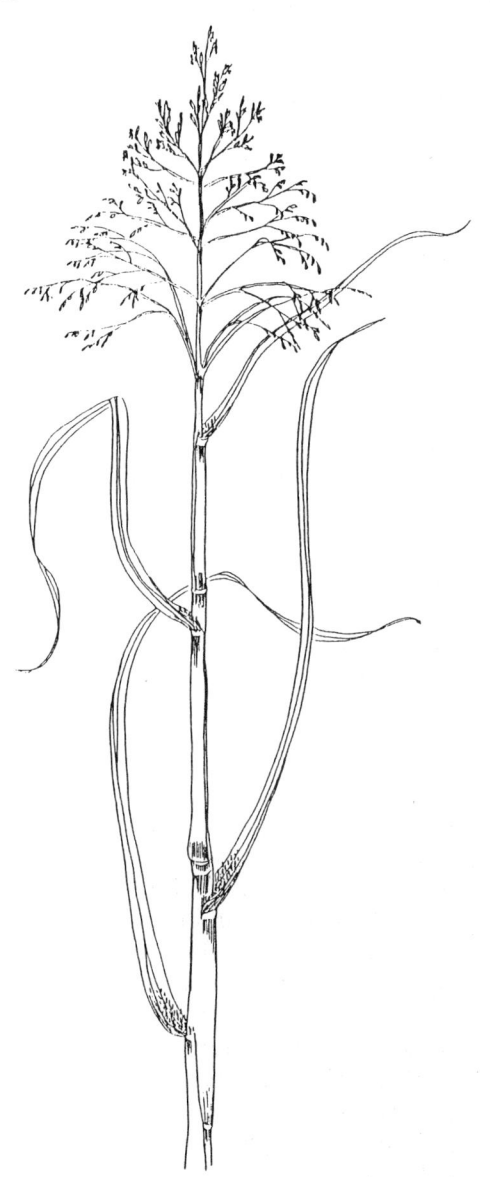

Sorghastrum nutans (L.) Nash, INDIAN GRASS

Height: 1–2 meters (3–6.2 feet)
Flowers: August–September
Color: Yellow anthers
Habitat: Mesic-Xeric

Identification Features:

The leaves of Indian Grass are rather stiff and spread at a 45 degree angle to the stem. They are lighter green and stiffer than those of Big Bluestem. The upper surface of the leaf blade possesses a prominent claw-like ligule where it joins the sheath. Showy yellow anthers protrude when the plant is in bloom. During fall the soft, silky-textured plumes become a beautiful golden color and gently sway with a breeze.

Ecological Notes:

This warm-season grass is a common associate of Big Bluestem. Both species have approximately the same growth habits, nutrient and moisture requirements. The root system of Indian Grass reaches depths of 1.6–1.8 m, extends laterally, and is slowly rhizomatous in the absence of tough competition. The decay of its deep, lateral fibrous root system is very important in soil formation. Indian Grass is relished by native and domestic ungulates. It also provides much fuel for prairie fires.

Spartina pectinata Link, PRAIRIE CORDGRASS (SLOUGHGRASS)

Height: 1–2.5 meters (3–8 feet)
Flowers: Mid July–August
Color: Yellow anthers
Habitat: Hydric

Identification Features:

The name "cordgrass" is suggestive of the toughness of its long, up to 80 cm, coarse leaves and thick, tough stems. The leaf blades have finely serrated margins that are quite sharp and capable of cutting skin. Showy yellow anthers protrude from the blooming florets. The seed spikes are up to 6 cm long. Cordgrass turns golden-yellow in the fall.

Ecological Notes:

This warm-season grass species occupies soils too wet and too poorly drained for the development of Big Bluestem and Switch Grass communities. Roots can penetrate to depths of 2.5–4 m into water-logged (poorly aerated) soil due to the large air-conducting space in their cortex. The upper 15–25 cm of soil beneath stands of Cordgrass is occupied by a mat of coarse, thick (5–10 mm), almost woody, highly branched rhizomes.

This sod-forming grass is aggressive in hydric soil conditions, often forming dense stands and excluding other species. In the center of solid stands of Cordgrass, grass height is much lower and there is rarely any flowering. Flowering usually occurs in plants around the edges. Plants in the center of dense stands store starch in their rhizomes to ensure survival rather than produce flowers or height. Cordgrass is relished by ungulates during spring and early summer before it becomes too tough.

Sporobolus heterolepis Gray, PRAIRIE DROPSEED

Height: 0.6–1.2 meters (2–4 feet)
Flowers: August
Color: Purplish stamens
Habitat: Mesic-Xeric

Identification Features:

Prairie Dropseed is a most attractive bunch-grass that grows in fountain-like tufts of 10–20 cm in diameter. Its slender leaves are 60 or more cm long and have a smooth texture. The inflorescence is a well-developed panicle about 20 cm long. It is fragrant while flowering. Following a burn, Prairie Dropseed bunches are noticeable as firm little mounds on the prairie soil.

Ecological Notes:

The presence of this warm-season grass in a prairie remnant indicates virgin conditions, as Prairie Dropseed does not tolerate heavy grazing or soil disturbances such as plowing. Its fibrous roots reach a depth of 1–1.7 m. The decay of these roots helps build the rich fertile soil of the grasslands. In some upland communities, it can be the dominant grass species. Prairie Dropseed is relished by ungulates and also provides fuel for prairie fires.

Stipa spartea Trin., NEEDLE (PORCUPINE) GRASS

Height: 0.5–1 meter (1.5–3 feet)
Flowers: June
Color: Inconspicuous
Habitat: Xeric

Identification Features:

Needle Grass gets its common name from the sharp, long twisting awns. This bunch-grass has long tapering leaves which are corrugated on the upper surface but smooth and shining beneath. Unlike other prairie grasses which turn a golden yellow or tan when mature, the leaves of Needle Grass become nearly white.

Ecological Notes:

This cool-season bunch-grass has strong, fibrous roots, 1–1.5 mm, that reach depths of 0.7–1 m. During periods of drought, the leaves of Needle Grass roll upward and inward to form a tube. The pores through which water vapor is normally transpired are now inside the tube while the exposed lower surface of the leaf is highly resistant to water loss.

Upon falling to the ground or coming into contact with an object, the sharp point of the seed penetrates by the twisting action of its awn. Needle Grass is very nutritious and relished by ungulates. However, it can also cause mechanical injury to livestock as the sharp point of the seed screws through fur and into the flesh by the twisting action of its awn.

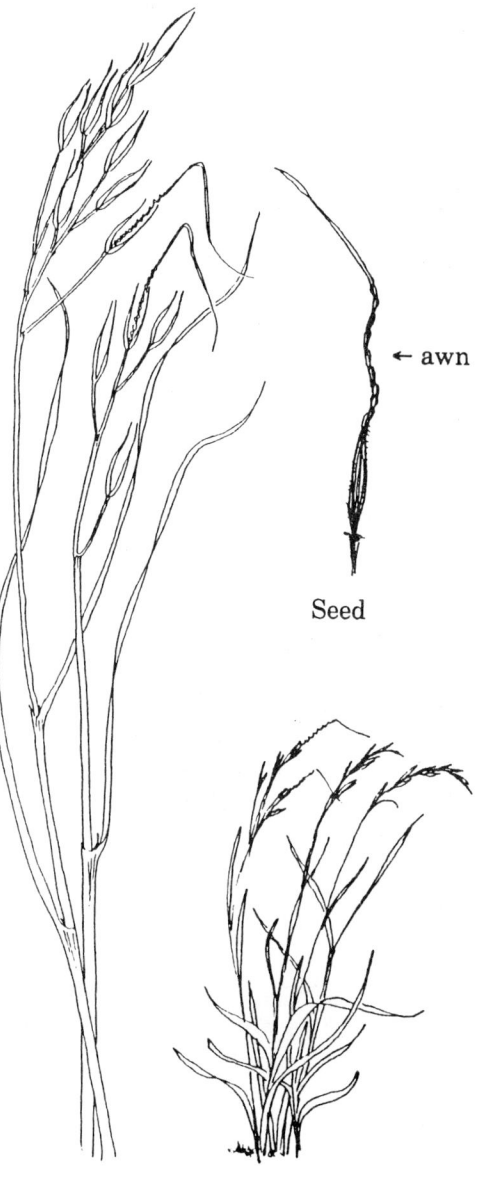

← awn

Seed

POLEMONIACEAE (PHLOX)

Phlox glaberrima interior Wherry, MARSH PHLOX

Height: 0.3–0.9 meter (1–3 feet)
Flowers: July–August
Color: Violet-Pink
Habitat: Hydric-Mesic

Identification Features:

The stem of Marsh Phlox is slender, erect, and smooth. Its opposite leaves are sessile, stiff, narrow and tapered. The inflorescence is a group of loosely branched clusters of violet-pink flowers that are on short pedicels. The flowers have a longer tube-like calyx than Prairie Phlox.

Ecological Notes:

Marsh Phlox thrives best in moist, calcareous prairies. It is pollinated by hummingbirds and long-tongued butterflies.

Phlox pilosa L., PRAIRIE PHLOX

Height: 0.3–0.7 meter (1–2 feet)
Flowers: Mid May–June
Color: Pink-Lavender
Habitat: Xeric-Mesic

Identification Features:

The stems of Prairie Phlox are slender, spreading, and hairy. Its opposite leaves are sessile, stiff, and slightly broader than those of Marsh Phlox. Large clusters of tight buds are furled in the shape of an umbrella. These clusters of buds develop into umbrella-like inflorescences of beautiful pink-lavender flowers. The flowers are on short pedicels and have a tube-like calyx.

Ecological Notes:

Prairie Phlox grows in prairie and Black Oak savanna. It is pollinated by long-tongued insects, such as butterflies.

POLYGALACEAE (MILKWORT)

Polygala senega L., SENECA SNAKEROOT

Height:	0.1–0.5 meter (4–20 inches)
Flowers:	May–Mid June
Color:	White
Habitat:	Xeric-Mesic

Identification Features:

Seneca Snakeroot has several stems in a clump growing from the root's thick crown which is just beneath the soil's surface. Its numerous deep-green leaves are alternate, sessile, and lanceolate to ovate. The upper leaves are longer than the lower ones. Dense racemes of very small, white flowers are at the tip of the stem. There are five sepals in the calyx. The uppermost and the two lower sepals are small and often greenish; the two lateral sepals, called "wings," are larger and white. Three petals are connected with each other and with the stamen tube. After the flowers fall, small bracts remain on the raceme.

Ecological Notes:

Seneca Snakeroot prefers calcareous soils. It is a member of stable prairie and savanna communities.

PRIMULACEAE (PRIMROSE)

Dodecatheon meadia L., SHOOTING STAR

Height:	0.2–0.6 meter (8–24 inches)
Flowers:	May–Early June
Color:	White-Pink-Lavender
Habitat:	Xeric-Mesic-Hydric

Identification Features:

One or more stiff, green stems of Shooting Star arise from a rosette of long (7–20 cm), strap-like, pale green basal leaves with reddish-pink midribs. The flower clusters from these upright stems are not crowded, each flower being on a long stalk. The flowers are uniquely shaped with their five re-curved petals, protruding pistil, and tightly pressed stamens pointing out. By July, the leaves have disappeared and only the stalk with its seed capsule remains.

Ecological Notes:

Shooting Star grows in prairie, savanna, rocky hillsides, and moist slopes. Pollination is by bumblebees.

Seed Capsules

RANUNCULACEAE (BUTTERCUP)

Anemone canadensis L., MEADOW ANEMONE

Height:	0.3–0.5 meter (1–1.5 feet)
Flowers:	June
Color:	White
Habitat:	Hydric-Mesic

Identification Features:

Meadow Anemone has hairy stems which arise from slender, tough rhizomes. Its leaves are five to seven parted with deep, irregular lobes and prominent veins. The basal leaves have long petioles and are found in whorls of three. The upper leaves are sessile and paired. Individual flowers bloom for only a short period of time. They have numerous stamens and pistils. The five white sepals are 2.5 to 4 cm long; there are no petals.

Ecological Notes:

Meadow Anemone forms dense colonies from the spread of its rhizomes. It is a common associate of Big Bluestem, especially in calcareous soils along moist roadside and railroad ditches. This habitat often occurs where gravel has spilled over the bank's edge. The showy flowers at the end of long stems are visible to bees and flies for pollination.

Anemone cylindrica Gray, THIMBLEWEED

Height:	0.3–0.7 meter (1–2 feet)
Flowers:	June–July
Color:	Greenish-White
Habitat:	Mesic-Xeric

Identification Features:

The tall, slender, fibrous stem of Thimbleweed arises from a bulbous rhizome. It has a whorl of deeply cut five-parted basal leaves with long petioles. Flowers arise from the tip of the top whorl of smaller leaves. The cylindrical receptacle has many pistils and stamens surrounding it. This receptacle is "thimble-shaped," thus giving this plant its common name. During fall, the thimble slowly disintegrates and its cottony seeds disperse into the air.

Ecological Notes:

Of all the Anemones, Thimbleweed has the greatest fidelity to prairie. The cottony seeds and fluff which remain until spring are sometimes gathered by hummingbirds and used to construct their miniature nests.

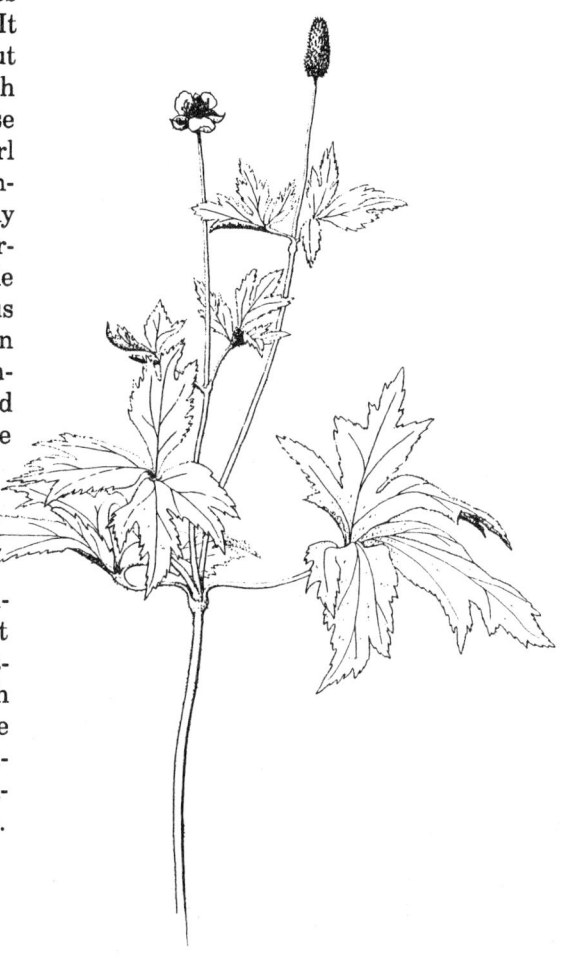

Anemone patens L., PASQUE FLOWER

Height:	0.1–0.3 meter (4–12 inches)
Flowers:	April
Color:	White to Lavender-Blue
Habitat:	Xeric

Identification Features:

Pasque Flower has gray-green stems which are covered with silky hairs. The stems grow from thick rhizomes and elongate after the flower blossoms. Each leaf has long silky hairs and is divided into three to seven deeply cut linear lobes. The basal leaves have long slender petioles, encircle the stem, and persist throughout the summer. The upper sessile leaves are smaller and occur just below the flower. A single large flower, resembling a crocus or tulip, grows from the top of each stem before the leaves appear. Each flower has five to seven sepals which are bluntly pointed. As with other *Anemone* species, there are no petals. Each seed has a long, fluffy plume. Pasque Flower is the earliest prairie wildflower to bloom!

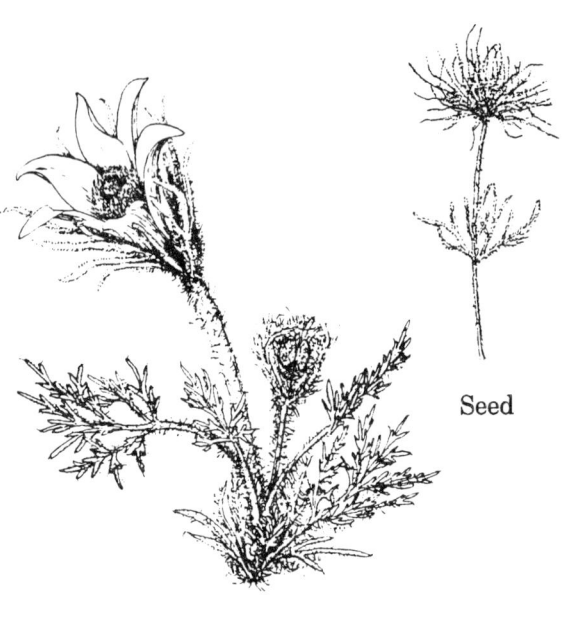

Seed

This species is highly variable and consists of several varieties. The one found growing east of the Mississippi River is *Anemone patens* L. var. *wolfgangiana* (Besser) Koch.

Ecological Notes:

Pasque Flower grows best in gravelly prairie and moraine hill habitats where there is little grass biomass. Because of its early growth habits, this species is harmed by spring fires and therefore thrives best in places where there is little to no fire.

RHAMNACEAE (BUCKTHORN)

Ceanothus americanus L., NEW JERSEY TEA

Height:	0.3–1 meter (1–3 feet)
Flowers:	Late June–July
Color:	White
Habitat:	Mesic

Identification Features:

This low, upright shrub has numerous slender stems arising from huge rootstocks. Clusters of white flowers arise from the tops of these stems. The dark green leaves of New Jersey Tea are alternate, finely toothed, veiny, oval, and stiff, with fine hairs beneath. As the seed capsules mature, they become black and are easily recognized. New Jersey Tea is a most attractive shrub at all stages during the growing season.

Ecological Notes:

Despite its common name, New Jersey Tea is found growing in prairies and savannas. The flowers are fragrant and attract many species of butterflies and other insects. New Jersey Tea is a food source for the larvae of the mottled dusky-wing butterfly, *Erynnis martialis*. Herbivores, especially deer and rabbits, browse heavily upon this shrub. New Jersey Tea can be burned or cut off to the ground while dormant; it will flower by early July on new wood. Its hard round seeds are stimulated to germinate by heat from fire.

ROSACEAE (ROSE)

Fragaria virginiana Duchesne, WILD STRAWBERRY

Height: 0.1–0.2 meter (4–8 inches)
Flowers: Mid April–Mid June
Color: White
Habitat: Xeric-Mesic

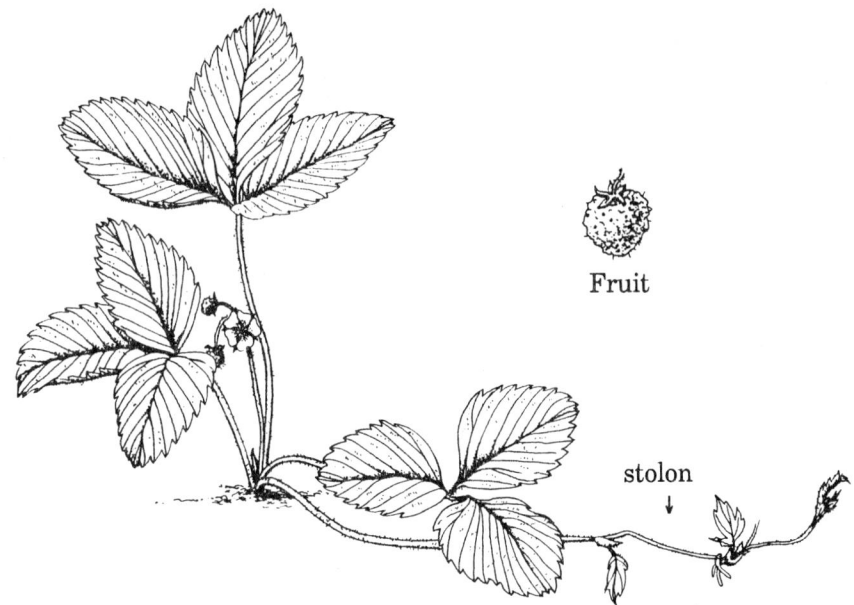

Fruit

stolon

Identification Features:

The horizontal stems of Wild Strawberry trail along the ground. Vegetative reproduction from these stolons forms open colonies. The leaves are on long pubescent petioles, up to 15 cm, which arise in tufts on the stolons. Each leaf is divided into three leaflets. The leaflets have serrated margins, are nearly sessile, and are pubescent on the underside. The flower has five petals and conspicuous yellow stamens. Wild Strawberry's fleshy, red globular fruit has tiny dry seeds, called achenes, in the minute pits on its surface.

Ecological Notes:

This species is not confined to prairie and savanna; it is also a component of old world meadow vegetation. Open colonies of Wild Strawberry are formed by vegetative reproduction from its stolons. The sweet and delicious fruits are prized by many herbivores and omnivores.

Geum triflorum Pursh, PRAIRIE SMOKE

Height: 0.15–0.4 meter (6–16 inches)
Flowers: May–Early June
Color: Pink-Purple
Habitat: Xeric

Identification Features:

The stem and basal leaves of Prairie Smoke grow from a rhizome at or near the soil's surface. Each flower stalk has three flowers at its summit. The basal leaves are 10–20 cm long and have 7–17 leaflets which are deeply cut and appear fern-like. Both the leaves and stem are pubescent. The flowers have five purplish petals which are inside the triangular sepals. The fruits have pinkish-purple, feathery tails that are up to 5 cm long. These fruits are the spectacular showy organs of Prairie Smoke.

Ecological Notes:

Prairie Smoke grows best in gravelly prairies, moraine hills, and savannas. These prairie habitats usually have little grass biomass. Its fruits are dispersed by wind. Because of its early growth habits, Prairie Smoke is harmed by spring fire and therefore thrives best in places where there is little to no fire.

Potentilla arguta Pursh, PRAIRIE CINQUEFOIL

Height:	0.5–0.8 meter (1.5–2.5 feet)
Flowers:	Late June–July
Color:	White
Habitat:	Mesic-Xeric

Identification Features:

The stout stem of Prairie Cinquefoil is highly pubescent. Its alternate leaves usually have seven to eleven pubescent leaflets. The leaflets have sharp coarse teeth along their margins. The flowers are in a rather dense cluster and resemble those of a strawberry. The tiny seeds of Prairie Cinquefoil are dry and enclosed within the infolded sepals.

Ecological Notes:

Prairie Cinquefoil is a member of the stable prairie.

Rosa carolina L., PASTURE ROSE

Height: 0.3–0.6 meter (1–2 feet)
Flowers: June–July
Color: Rose-Pink
Habitat: Xeric-Mesic

Identification Features:

Pasture Rose is one of the few true prairie shrubs. New shoots arise from its spreading root system. The stems have slender, straight, weak prickles. The alternate leaves are divided into five to eleven oblong leaflets with serrated margins. The fragrant flowers are often on new wood; they may be solitary or in clusters. They have five showy petals, five green sepals, and many bright yellow stamens in the center. The bright red fruit, known as a "hip," has hard seeds inside the fleshy, bright red receptacle.

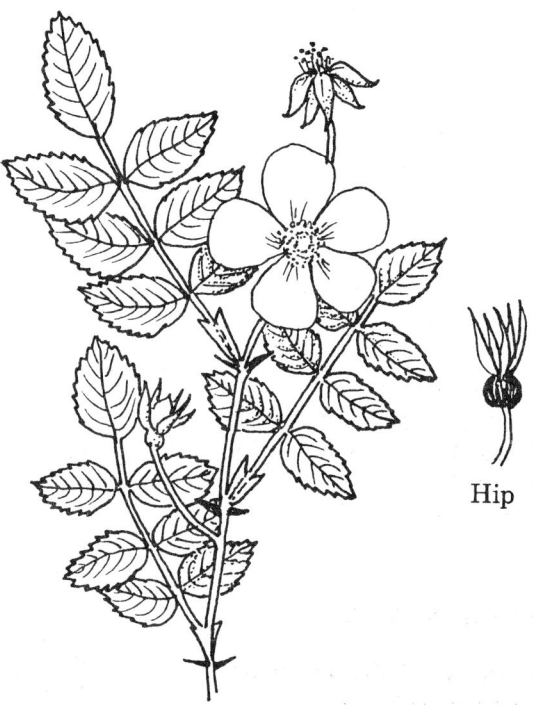

Hip

Ecological Notes:

Pasture Rose is one of the few prairie plants that are truly woody, most are herbaceous. This species is common in eastern prairies. During the winter, its fruit provides valuable food for wildlife such as deer, pheasants, and small mammals. Honeybees are common pollinators of wild roses.

Generally, rose species of the prairie have weak prickles. There was little selection pressure for the development of stout prickles in true prairie species of this genus because aggressive grazing ungulates, such as cattle and sheep, were absent. One related species, *Rosa blanda* Ait., EARLY WILD ROSE, is almost devoid of prickles.

RUBIACEAE (MADDER)

Galium boreale L., NORTHERN BEDSTRAW

Height:	0.2–0.6 meter (8–24 inches)
Flowers:	June–Mid July
Color:	White
Habitat:	Mesic-Hydric

Identification Features:

The fine, square stems of Northern Bedstraw are hollow, weak and sprawling. The slender leaves are tapered at the end and are in whorls of four. They are strongly three-nerved with a prominent midrib and rough margin. Clusters of white, tiny, saucer-shaped flowers are borne in the upper leaf axils. Each flower has four tiny lobes. The black fruit is smooth to slightly bristly.

Ecological Notes:

Northern Bedstraw grows in dense colonies.

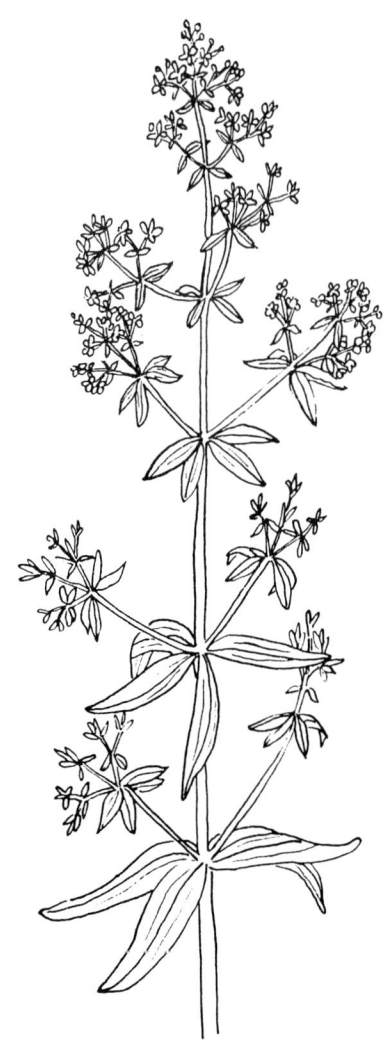

SALICACEAE (WILLOW)

Salix humilis Marshall, PRAIRIE WILLOW

Height:	1–2.2 meters (3–8 feet)
Flowers:	Mid April–Mid May
Color:	Gray-White
Habitat:	Xeric

Identification Features:

Prairie Willow has brownish-yellow woody stems which are slightly pubescent. Its alternate leaves are lanceolate, 5–15 cm long, 1–3 cm wide, and taper to each end. The leaves expand after the fruit matures and become broadest near the middle. Their upper surface is dark green while the lower surface is gray and pubescent. The flowering spikes, called catkins, appear before the leaves.

This species is dioecious; the staminate and pistillate flowers are on separate colonies. The pistillate catkins are 2.5–3 cm long. The staminate catkins are shorter and reddish-tinged.

Prairie Willow is a highly variable species.

Ecological Notes:

Prairie Willow is one of the few true woody plants of the prairie; most prairie plants are herbaceous. This species forms colonies by rhizomes and is rarely found in disturbed sites.

SANTALACEAE (SANDALWOOD)

Comandra umbellata (L.) Nutt., FALSE TOADFLAX

Height:	0.1–0.3 meter (4–12 inches)
Flowers:	May–June
Color:	Creamy-White
Habitat:	Xeric

Identification Features:

False Toadflax has smooth, erect stems which arise from creeping rhizomes growing just under the soil's surface. Its root system is poorly developed. The numerous leaves are alternate, sessile, oblong, 2–4 cm long, thick and leathery. They have a prominent midrib on the underside. The bell-shaped flowers occur in small clusters at the tips of the stem. The flowers have five creamy-white sepals; there are no petals.

Ecological Notes:

False Toadflax grows in a wide range of habitats. It is partially parasitic on the roots of other plants. False Toadflax spreads by rhizomes and forms dense colonies. It survives abuse, such as overgrazing.

SAXIFRAGACEAE (SAXIFRAGE)

Heuchera richardsonii R. Br., PRAIRIE ALUM ROOT

Height:	0.3–1 meter (1–3 feet)
Flowers:	June–Early July
Color:	Orange-Creamish-Brown-Green
Habitat:	Xeric-Mesic-Hydric

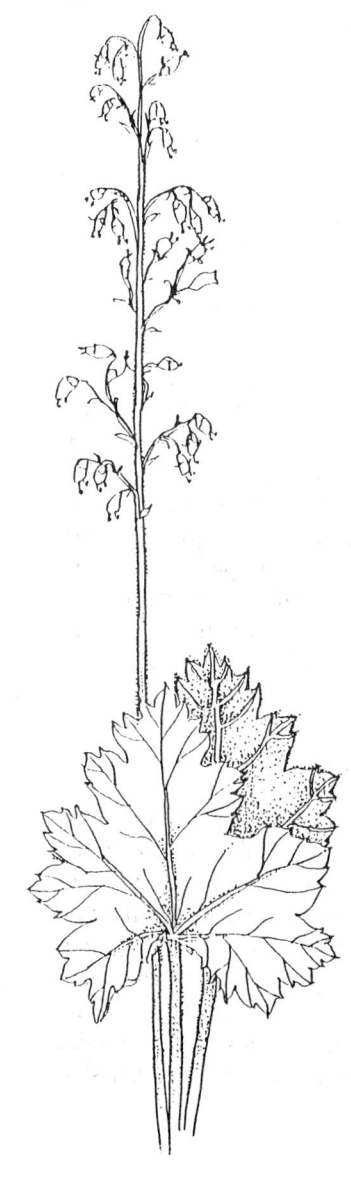

Identification Features:

The flowering stem of Prairie Alum Root arises from a cluster of geranium-like leaves. This broad cluster of scalloped, round, hairy leaves is handsome throughout the growing season. The flowers are small, 3–5 mm, and bell-like, with beautiful colorful anthers protruding out on green filaments.

This species is variable and consists of different varieties. The ones found growing east of the Mississippi River are varieties *affinis* and *grayana*.

Ecological Notes:

Prairie Alum Root is a member of prairie and savanna communities. Aphids often attach themselves to the small attractive flowers.

SCROPHULARIACEAE (FIGWORT)

Castilleja coccinea (L.) Spreng, INDIAN PAINT BRUSH

Height:	0.15–0.6 meter (6–24 inches)
Flowers:	Late May–June
Color:	Scarlet-Orange-Yellow
Habitat:	Xeric-Mesic

Flower

Identification Features:

Indian Paint Brush has several pubescent stems which grow from a single base. Its leaves are alternate and sessile. The basal leaves are grass-like, 2–5 cm long and very narrow. The upper leaves, called bracts, are commonly three-lobed. These bracts are the spectacular showy organs of this species! Their color varies from bright scarlet to orange-yellow. The small, two-lipped yellowish flowers and their protruding pistils are almost hidden by the brightly colored bracts.

Ecological Notes:

The long and narrow yellow flowers of Indian Paint Brush are highly adapted for insect pollination. The roots are partially parasitic on the roots of other plants. Unlike most prairie plant species, Indian Paint Brush is not a perennial; its life cycle can be biennial or annual.

Pedicularis canadensis L., WOOD BETONY (LOUSEWORT)

Height: 0.15–0.4 meter (6–16 inches)
Flowers: May–June
Color: Yellow
Habitat: Xeric-Mesic

Identification Features:

Wood Betony has tufted pubescent stems which grow from rhizomes. The leaves are alternate, mainly basal, up to 13 cm long, thick, and pubescent. They are pinnately lobed which gives them a fern-like appearance. In early spring, the basal leaves are reddish before they turn green. The snapdragon-shaped flowers are strongly two-lipped; two of the five petals are fused to form an arched hood while the other three are fused to form the lower lip. The flowers are in dense clusters toward the summit of the stem.

Ecological Notes:

This species prefers acidic soils of prairie and oak savanna. Lousewort is partially parasitic on the roots of other plants, especially grasses. The flowers are pollinated by bees.

Penstemon digitalis Nutt., FOXGLOVE BEARD TONGUE

Height: 0.3–1.0 meter (1–3 feet)
Flowers: June–Early July
Color: White
Habitat: Xeric

Identification Features:

Foxglove Beard Tongue has a stiff, smooth stem that arises from an enlargement of the taproot. The waxy leaves are opposite, thick, toothed, and sessile. Flower stalks grow in pairs from the upper leaf axils. White, tubular "two-lipped" flowers are scattered along the upper portion of the stem. The white corolla has an upper two-lobed lip and a lower three-cleft lip. Each flower has a tufted, hairy sterile stamen which resembles a "beard."

This species is barely distinguishable from *Penstemon calycosus* Small, SMOOTH BEARD TONGUE, which has purple-tinged corollas.

Ecological Notes:

This species prefers calcareous rocky slopes. The thick, glossy, waxy leaves of Foxglove Beard Tongue help to minimize the loss of moisture from the leaves by evapotranspiration. Most pollination is accomplished by insects.

Veronicastrum virginicum (L.) Farw., CULVER'S ROOT

Height:	0.6–1.6 meters (2–5 feet)
Flowers:	July–August
Color:	White
Habitat:	Hydric-Mesic

Identification Features:

The tall, erect stems of Culver's Root have leaves whorled around each of their nodes. Its leaves are 7–20 cm long and 2–2.5 cm wide and have slightly saw-toothed edges. The tiny, white flowers are about 2 mm long. They are densely crowded on several spike-like racemes up to 20 cm long. The racemes may have several branches.

Ecological Notes:

Culver's Root is a member of prairie and savanna communities. The presence of this species indicates good soil and moisture conditions. It has a fragrant scent and is visited by many insects such as small beetles and sulfur butterflies.

VIOLACEAE (VIOLET)

Viola pedatifida G. Don, PRAIRIE VIOLET

Height: 0.1–0.25 meter (4–10 inches)
Flowers: May
Color: Blue-Purple
Habitat: Mesic

Identification Features:

The flower stalk and leaves of Prairie Violet grow from a crown, developed from a rhizome, just beneath the soil's surface. The leaves are divided into three, then divided again into two to four lobes. These repeatedly divided lobes are long and narrow. Blue-purple flowers rise above the foliage on long stalks which curve downward at the top. Each flower has five blue-purple petals. The three lower petals are white at the base and have hairy beards on the inner surfaces. As with all *Viola* spp., the lowest petal has a sac-like base called a "spur."

Cleistogamous Flower

Ecological Notes:

When the seeds of Prairie Violet are ripe, the capsule opens and ejects them. In addition to its blue-purple regular flowers, this species also bears cleistogamous flowers on ascending stalks. Cleistogamy refers to the self-fertilization of small flowers which never open. Cleistogamous flowers have vestigial petals, no nectar, and remain in a green, bud-like stage. They become noticeable after the regular flowers have disappeared. If the regular flowers are not fertilized or do not develop, cleistogamous flowers help to ensure a next generation.

SELECTED REFERENCES

Anderson, Roger C. 1991. Illinois prairies: A historical perspective. Pages 384–391. *In* L. M. Page and M. R. Jeffords (eds.) *Our Living Heritage: The Biological Resources of Illinois.* Bulletin 34(4): 357–477. Illinois Natural History Survey, Champaign, Illinois.

Axelrod, Daniel L. 1985. Rise of the grassland biome, Central North America. *The Botanical Review* 51(2): 163–201.

Betz, Robert F. and Herbert F. Lamp. 1990. Flower, pod, and seed production in eighteen species of milkweeds (*Asclepias*). Pages 25–30. *In* D. D. Smith and C. A. Jacobs (eds.) *Proceedings of the Twelfth North American Prairie Conference.* University of Northern Iowa, Cedar Falls, Iowa.

Cruden, R. W., L. Hermanutz, and J. Shuttleworth. 1984. The pollination biology and breeding system of *Monarda fistulosa* (Labiatae). *Oecologia* 64: 104–110.

Farrar, Jon. 1990. *Field Guide to Wildflowers of Nebraska and the Great Plains.* Nebraskaland Magazine, Nebraska Game and Parks Commission, Lincoln, Nebraska.

Fernald, Merritt L. 1950. *Gray's Manual of Botany*, 8th Edition. Dioscorides Press, Portland, Oregon.

Frost, S. W. 1945. Insects feeding or breeding on indigo, *Baptisia. Journal New York Entomological Society*, 53: 219–225.

Gleason, Henry Allan. 1974. *The New Britton and Brown Illustrated Flora of the Northeastern United States and Adjacent Canada.* 5th Printing. New York: New York Botanical Garden.

Gleason, Henry A. and Arthur Cronquist. 1963. *Manual of Vascular Plants of Northeastern United States and Adjacent Canada.* D. Van Nostrand Company, Inc., Princeton, New Jersey.

Gleason, Henry A. and Arthur Cronquist. 1991. *Manual of Vascular Plants of Northeastern United States and Adjacent Canada.* 2nd Edition. The New York Botanical Garden, Bronx, New York.

Haddock, R. C. and S. J. Chaplin. 1982. Pollination and seed production in two phenologically divergent prairie legumes (*Baptisia leucophaea* and *B. leucantha*). *The American Midland Naturalist*, 108: 175–186.

Johnson, James R. and James T. Nichols. 1970. *Plants of South Dakota Grasslands*, South Dakota State University, Brookings, South Dakota.

Kerster, H. W. 1968. Population age structure in the prairie forb, *Liatris aspera*. *Bioscience* 18(5): 430–432.

Küchler, A. W. Potential natural vegetation of the conterminous United States (a map). American Geographical Society, New York.

Madson, John. 1982. *Where the Sky Began: Land of the Tallgrass Prairie*. Sierra Club Books, San Francisco.

McKone, Mark J. and David D. Biesboer. 1986. Nitrogen fixation in association with the root systems of goldenrods (*Solidago* L.). *Soil Biology & Biochemistry*, 18(5): 543–545.

Orwig, Timothy T. 1990. Loess Hills prairies as butterfly survivia: opportunities and challenges. Pages 131–135. *In* D. D. Smith and C. A. Jacobs (eds.) *Proceedings of the Twelfth North American Prairie Conference*. University of Northern Iowa, Cedar Falls, Iowa.

Owensby, Clenton E. 1980. *Kansas Prairie Wildflowers*. Iowa State University Press, Ames, Iowa.

Runkel, Sylvan T. and Dean M. Roosa. 1989. *Wildflowers of the Tallgrass Prairie: The Upper Midwest*. Iowa State University Press, Ames, Iowa.

Samson, Fred and Fritz Knopf. 1994. Prairie conservation in North America. *Bioscience* 44(6): 418–421.

Schaal, Barbara A. 1978. Density dependent foraging on *Liatris pycnostachya*. *Evolution* 32(2): 452–454.

Schlicht, Dennis W. and Timothy T. Orwig. 1990. Sequential use of niche by prairie obligate skipper butterflies (Lepidoptera: Hesperidae) with implications for management. Pages 137–139. *In* D. D. Smith and C. A. Jacobs (eds.) *Proceedings of the Twelfth North American Prairie Conference*. University of Northern Iowa, Cedar Falls, Iowa.

Smith, Helen V. 1961. *Michigan Wildflowers*. Cranbrook Institute of Science, Bloomfield Hills, Michigan.

Swink, Floyd and Gerould Wilhelm. 1979. *Plants of the Chicago Region*, 3rd Edition. The Morton Arboretum, Lisle, Illinois.

Swink, Floyd and Gerould Wilhelm. 1994. *Plants of the Chicago Region*, 4th Edition. Indiana Academy of Science, Indianapolis, Indiana.

Transeau, Edgar N. 1935. The prairie peninsula. *Ecology* 16(3): 423–437.

Van Bruggen, Theodore. 1971. *Wildflowers, Grasses & Other Plants of the Northern Plains and Black Hills*. Badlands Natural History Association, Interior, South Dakota.

Voight, John W. and Robert H. Mohlenbrock. 1978. *Prairie Plants of Illinois*, Department of Conservation, Springfield, Illinois.

Voss, John and Virginia S. Eifert. 1951. *Illinois Wild Flowers*, Department of Registration and Education, Springfield, Illinois.

Weaver, John E. 1968. *Prairie Plants and Their Environment: A Fifty-Year Study in the Midwest*. University of Nebraska Press, Lincoln, Nebraska.

Weaver, John E. 1954. *North American Prairie*, Johnsen Publishing Company, Lincoln, Nebraska.

Werner, Patricia A. 1978. On the determination of age in *Liatris aspera* using cross-sections of corms: Implications for past demographic studies. *The American Naturalist*, 112(988): 1113–1119.

GLOSSARY

Abiotic. Non-living components of the environment, such as bedrock, water, air, light, and nutrients.

Abscission layer. A layer of special cells at the base of a plant part, such as the stem, that allows it to separate from the rest of the plant.

Achene. A small, dry, hard, one-seeded fruit.

Alkaline. Material that is basic rather than acidic; the pH is greater than 7.0.

Allelopathic. Refers to the suppression or destruction of plants by toxic chemicals produced and released by other, nearby plants.

Anatomical. Referring to the structural make-up of an organism or any of its parts.

Annual. A plant which germinates from seed, flowers, produces seed, and dies in one growing season.

Anther. The pollen-bearing part of the stamen.

Awn. A bristle-like appendage, common on grasses, attached to a lemma or glume.

Axil. The angle between two plant parts, such as the stem and leaf.

Basal. Positioned at, or arising from, the base of the plant or some organ of the plant.

Biennial. A plant that normally lives two years. During the first year, a biennial forms a basal rosette of leaves; during the second year, it flowers, produces fruit, and dies.

Biomass. The amount of organic matter per unit area or volume.

Biome. A broad vegetational land region, characterized by a similar climate, soil, topography, plants and animals, regardless of where it occurs on Earth. For example, tropical rain forest, desert, and grassland.

Blade. The flattened expanded portion of a leaf.

Bract. A reduced or otherwise modified leaf which subtends a flower or flower cluster.

Bulb. Usually an underground leaf bud, consisting of a thickened short stem that is overlapped with scalelike leaves, as in lily or onion.

Bulblet. A small bulb, usually borne in the leaf axil.

Calcareous. Of, or pertaining to, a high concentration of calcium carbonate (limestone) in the soil.

Calyx. The outer whorl (usually green) of flower parts; collective for sepals.

Cordate. A heart-shaped leaf with the notch at the base.

Corm. A solid bulblike stem that is usually underground.

Corolla. The inner whorl (usually colored) of flower parts; collective term for the petals.

Corona. A crown of petal-like structures between the petals and stamens.

Cylindric. Elongated with a circular cross section.

Dioecious. The staminate and pistillate flowers are on different plants.

Elliptic. A circular shape which is broadest at the middle.

Epidermis. The covering tissue of roots, leaves, and stems of nonwoody plants.

Evapotranspiration. The loss of water from plants.

Fen. Wetland communities where carbonate-rich water discharges at a constant rate along the slopes of moraines, bluffs, or other glacial formations. The water has a high content of calcium or magnesium carbonates.

Filament. The threadlike stalk of the stamen, which supports the anther.

Floret. A very small flower, particularly one found in a dense inflorescence, as in the composites or in the grasses.

Forb. A herbaceous plant other than grass, sedge, or rush.

Fungi. Organisms, comprised of molds, mildews, rusts, smuts, and mushrooms, that lack chlorophyll and reproduce mainly by means of asexual spores.

Gland. A small protuberance, or other structure, which secretes sticky or oily substances.

Glume. One of the two empty chafflike bracts at the base of a grass spikelet.

Head. A dense cluster of sessile or subsessile flowers; as in whorls of bracts subtending the inflorescence of composites.

Herbivore. An organism that eats plants and plant products.

Hydric. Wet, especially relating to the soil moisture conditions.

Inflorescence. A flower cluster.

Internode. The portion of the stem between two successive nodes.

Involucral. Referring to the involucre which are whorls of bracts subtending an inflorescence; as around the base of a composite head.

Lanceolate. Lance-shaped; long tapering above the middle and several times longer than wide.

Leaflet. A division of a compound leaf. Leaflets resemble leaves but do not have buds in their axils.

Lemma. The lower of two scale-like bracts that enclose a grass flower or seed; located directly above the glumes.

Lignified. "Woody" as a result of conversion of compounds of the cell wall into lignin.

Ligule. A small strap-shaped appendage at the junction of the blade and sheath of a grass leaf; also, one of the strap-shaped corollas of the composites.

Loment. A flat legume fruit that is conspicuously constricted between the seeds, falling apart into one-seeded sections at the constrictions when mature.

Mesic. Medium, especially relating to the soil moisture conditions.

Midrib. The central rib or vein of a leaf or other organ.

Moraine. A mass of rocks, gravel, sand, clay or some other earth material, carried and deposited by a glacier.

Mucilaginous. Slimy, gelatinous, carbohydrate substances, as in sap.

Nectar. A sweet liquid rich in sugars, amino acids, and other compounds that is produced by various plant parts such as the stigma and nectar glands.

Node. The point on a stem where leaves, branches, or inflorescences arise.

Ovate. Egg-shaped, the broadest part below the middle.

Palmate. Having parts deeply lobed and diverging from a common base.

Palatability. A quality of forage that is pleasing and acceptable to the taste of a grazing animal.

Pappus. Outgrowths on the achenes of composites consisting of bristles, hairs, scales or awns.

Panicle. A cluster of flowers arranged in a series of two or more racemes.

Pedicel. The stalk of a single flower in an inflorescence.

Perennial. A plant whose life cycle is longer than two years, not a biennial.

Perigynium. An inflated sac enclosing the pistil, as in the genus *Carex*.

Petal. A part of the corolla.

Petiole. The stalk of a leaf.

pH. The relative concentration of H^+ ions in a solution. Low pH values (less than 7) indicate high concentrations of H^+ ions (acidic), and high pH values (greater than 7) indicate low concentrations of H^+ ions (alkaline or basic).

Pistil. The female reproductive structure of a flower consisting of the stigma, style, and ovary.

Pistillate. Refers to plants, flowers, or inflorescences which bear pistils but not stamens.

Predation. The relationship in which one organism (predator) consumes another organism (prey) for nutrition.

Prickle. A sharp, usually small or slender, outgrowth of the epidermis or bark.

Prostrate. Lying flat on the ground.

Pubescent. Having the surface covered with soft hairs.

Raceme. An inflorescence whose stalked flowers are arranged along an elongated axis.

Ray. A strap-shaped (ligulate) marginal flower in the head of the composite inflorescence.

Receptacle. The floral axis to which the various flower parts are attached; for example, the disk or dome-shaped structure of the composite family that bears the florets.

Resin. A semi-solid, viscous, sticky organic substance exuded from a plant.

Resinous. To secrete or exude resin.

Rhizome. An underground, horizontal stem which can produce shoots and roots at the nodes, giving rise to new plants.

Rootstock. An underground stem; used to designate a rhizome or rhizome-like structure.

Rosette. A cluster of leaves usually in a circular arrangement at the base of a plant.

Savanna. Grassland communities that have scattered trees. The tree canopy does not totally shade the ground surface.

Sepal. A part of the calyx.

Sessile. Without a stalk of any kind.

Sheath. A tubular structure, consisting of the lower part of the leaf, which clasps or encloses the stem, especially in grasses and sedges.

Spike. An unbranched inflorescence in which the flowers are sessile or sub-sessile on a central axis.

Spikelet. A small or secondary spike. In sedges and grasses, it is the flower cluster consisting of one to many flowers subtended by two bracts (the bracts are known as glumes in grasses).

Spur. An extended sac at the base of a corolla. Also, a short shoot (branchlet) with a very compact arrangement of leaf scars.

Stalk. The stem of any organ.

Stamen. The pollen-bearing organ of the flower, composed of the anther and filament.

Staminate. Refers to plants, flowers, or inflorescences which bear stamens but not pistils.

Stigma. The part of a pistil or style that receives the pollen.

Stipule. A leaf-like appendage found at the point of attachment of a leaf to the stem; usually occur in pairs.

Stolon. A modified, above-ground, horizontal stem that roots at the nodes and produces new plants.

Succulent. Thickened leaves, stems, or roots that are juicy and fleshy.

Swale. A low, shallow, drainage area of land that is often wet.

Symbiotic. Living together, referring to two dissimilar organisms, with benefit, to one or both, but without harm to either.

Taproot. The primary descending root, usually thickened and larger than others in the root system.

Tendril. A slender, twining organ used for support or climbing.

Transpiration. The loss of water from the plant to the atmosphere, occurring mainly through evaporation at leaf stomata.

Tuber. An enlarged, thickened, fleshy portion of an underground stem, usually functioning as a food reserve organ.

Umbel. A flat-topped or convex inflorescence in which the branches (pedicels) all radiate from a common point.

Ungulate. A hoofed mammal, such as a buffalo.

Whorls. A ring-like arrangement of three or more organs arising from a common point or node.

Xeric. Dry, especially relating to soil moisture conditions.

FAMILY NAME INDEX

Apiaceae (Carrot or Parsley)	1	Orchidaceae (Orchid)	80
Asclepiadaceae (Milkweed)	5	Oxalidaceae (Oxalis)	83
Asteraceae (Composite or Sunflower)	9	Poaceae (Grass)	84
Boraginaceae (Borage)	42	Polemoniaceae (Phlox)	100
Cactaceae (Cactus)	43	Polygalaceae (Milkwort)	102
Commelinaceae (Spiderwort)	44	Primulaceae (Primrose)	103
Cyperaceae (Sedge)	45	Ranunculaceae (Buttercup)	104
Euphorbiaceae (Spurge)	49	Rhamnaceae (Buckthorn)	107
Fabaceae (Legume)	50	Rosaceae (Rose)	108
Gentianaceae (Gentian)	64	Rubiaceae (Madder)	112
Iridaceae (Iris)	67	Salicaceae (Willow)	113
Lamiaceae (Mint)	69	Santalaceae (Sandalwood)	114
Liliaceae (Lily)	72	Saxifragaceae (Saxifrage)	115
Linaceae (Flax)	77	Scrophulariaceae (Figwort)	116
Lobeliaceae (Lobelia)	78	Violaceae (Violet)	120
Onagraceae (Evening Primrose)	79		

INDEX TO LATIN NAMES

Allium cernuum	72	*Camassia scilloides*	73
Amorpha canescens	51	*Carex*	
		bicknellii	45
Andropogon		meadii	46
gerardii	86	pensylvanica	47
scoparius	87	stricta	48
		tetanica	46
Anemone			
canadensis	104	*Castilleja coccinea*	116
cylindrica	105		
patens	106	*Ceanothus americanus*	107
patens wolfgangiana	106		
		Cirsium	
Asclepias	5	discolor	16
hirtella	6	hillii	16
sullivantii	7		
tuberosa	8	*Comandra umbellata*	114
Aster	10	*Coreopsis*	
azureus	10	palmata	17
ericoides	11	tripteris	18
laevis	12		
novae-angliae	13	*Cypripedium candidum*	80
sericeus	14		
		Dalea	
Astragalus canadensis	52	candida	60
		purpurea	61
Baptisia	53		
leucantha	55	*Desmodium*	57
leucophaea	56	canadense	57
		illinoense	58
Bouteloua curtipendula	88		
		Dodecatheon meadia	103
Bromus kalmii	89		
		Echinacea pallida	19
Cacalia			
plantaginea	15	*Elymus canadensis*	91
tuberosa	15		
		Eryngium yuccifolium	1
Calamagrostis canadensis	90		
		Euphorbia corollata	49

Fragaria virginiana	108	*Lilium*	
Galium boreale	112	michiganense	75
		philadelphicum andinum	76
Gentiana		tigrinum	75
andrewsii	64	*Linum sulcatum*	77
flavida	65		
puberulenta	66	*Lithospermum*	
		canescens	42
Geum triflorum	109	croceum	42
Habenaria leucophaea	81	*Lobelia spicata*	78
Helianthus	20	*Monarda fistulosa*	69
laetiflorus rigidus	22		
mollis	20	*Oenothera pilosella*	79
occidentalis	21		
rigidus	22	*Opuntia humifusa*	43
Heliopsis helianthoides	23	*Oxalis violacea*	83
Heuchera	115	*Panicum*	
richardsonii	115	leibergii	94
richardsonii affinis	115	oligosanthes scribnerianum	94
richardsonii grayana	115	virgatum	95
Hierochloe odorata	92	*Parthenium integrifolium*	28
Hypoxis hirsuta	74	*Pedicularis canadensis*	117
Iris virginica shrevei	67	*Penstemon*	
		calycosus	118
Koeleria cristata	93	digitalis	118
Kuhnia eupatorioides	24	*Petalostemum*	60
		candidum	60
Lespedeza capitata	59	purpureum	61
Liatris	25	*Phlox*	
aspera	26	glaberrima interior	100
pycnostachya	27	pilosa	101
spicata	27	*Physostegia virginiana*	70

Polygala senega	102
Polytaenia nuttallii	2
Potentilla arguta	110
Prenanthes	
aspera	29
racemosa	29
Psoralea tenuiflora	62
Pycnanthemum virginianum	71
Ratibida pinnata	30
Rosa	
blanda	111
carolina	111
Rudbeckia hirta	31
Salix humilis	113
Senecio pauperculus	32
Schizachyrium scoparium	87
Silphium	33
integrifolium	34
laciniatum	35
perfoliatum	36
terebinthinaceum	37
Sisyrinchium albidum	68
Solidago	38
missouriensis fasciculata	38
nemoralis	39
rigida	40
Sorghastrum nutans	96
Spartina pectinata	97

Sporobolus heterolepis	98
Stipa spartea	99
Tradescantia ohiensis	44
Vernonia fasciculata	41
Veronicastrum virginicum	119
Vicia americana	63
Viola pedatifida	120
Zizia	
aptera	3
aurea	4

INDEX TO COMMON NAMES

Alum Root	115	Compass Plant	35
American Vetch	63	Coneflower	
Aster	10	Gray-headed	30
Heath	11	Purple	19
New England	13	Yellow	30
Silky	14	Coreopsis	
Sky-blue	10	Prairie	17
Smooth Blue	12	Tall	18
		Cream Gentian	65
Balsam Ragwort	32	Cream Wild Indigo	56
Beard Tongue		Culver's Root	119
Foxglove	118	Cup Plant	36
Smooth	118		
Beebalm	69	Downy Gentian	66
Bicknell's Sedge	45	Downy Sunflower	20
Big Bluestem	86	Dropseed	98
Black-eyed Susan	31		
Blazing Star	25	Early Wild Rose	111
Button	26		
Marsh Blazing	27	False Boneset	24
Prairie	27	False Dragonhead	70
Rough	26	False Sunflower	23
Blue Flag Iris	67	False Toadflax	114
Blue-eyed Grass	68	Flowering Spurge	49
Bluestem		Foxglove Beard Tongue	118
Big	86		
Little	87	Gentian	
Blue Joint Grass	90	Bottle	64
Bottle Gentian	64	Closed	64
Butterfly Milkweed	8	Cream	65
Button Blazing Star	26	Downy	66
		Prairie	66
Canada Tick Trefoil	57	Yellowish	65
Canada Wild Rye	91	Glaucous White Lettuce	29
Canadian Milk Vetch	52	Golden Alexanders	4
Closed Gentian	64	Goldenrod	38
Clover	60	Missouri	38
Purple Prairie	61	Old-field	39
Round-headed Bush	59	Rigid	40
White Prairie	60	Stiff	40
Common Ironweed	41	Gray-headed Coneflower	30
Common Mountain Mint	71		

Grass	84
Blue-eyed	68
Blue Joint	90
Indian	96
June	93
Needle	99
Porcupine	99
Prairie Panic	94
Scribner's Panic	94
Switch	95
Vanilla	92
Yellow Star	74
Grooved Yellow Flax	77
Hairy Puccoon	42
Heart-leaved Meadow Parsnip	3
Heath Aster	11
Hoary Puccoon	42
Illinois Tick Trefoil	58
Indian Grass	96
Indian Paint Brush	116
Indian Plantain	15
Indigo	53
Cream Wild	56
White Wild	55
June Grass	93
Lead Plant	51
Lily	
Prairie	76
Tiger	75
Turk's Cap	75
Little Bluestem	87
Lousewort	117
Marsh Blazing Star	27
Marsh Phlox	100
Meadow Anemone	104
Mead's Sedge	46
Milkweed	5
Butterfly	8
Common	7

Prairie	7
Tall Green	6
Missouri Goldenrod	38
Needle Grass	99
New England Aster	13
New Jersey Tea	107
Nodding Wild Onion	72
Northern Bedstraw	112
Oak Sedge	47
Obedient Plant	70
Old-field Goldenrod	39
Pale Spiked Lobelia	78
Panic Grass	
Scribner's	94
Prairie	94
Pasque Flower	106
Pasture Rose	111
Pasture Thistle	16
Pennsylvania Sedge	47
Phlox	
Prairie	101
Marsh	100
Pleurisy Root	8
Porcupine Grass	99
Prairie	
Alum Root	115
Blazing Star	27
Brome	89
Cinquefoil	110
Cordgrass	97
Coreopsis	17
Dock	37
Dropseed	98
Gentian	66
Lily	76
Milkweed	7
Parsley	2
Panic Grass	94
Phlox	101
Puccoon	42
Sedge	45

Prairie *(continued)*	
Smoke	109
Sundrops	79
Sunflower	22
Thistle	16
Violet	120
White Fringed Orchid	81
Willow	113
Prickly Pear	43
Puccoon	
Hairy	42
Hoary	42
Prairie	42
Purple Coneflower	19
Purple Prairie Clover	61
Ragwort	
Balsam	32
Prairie	32
Rattlesnake Master	1
Rigid Goldenrod	40
Rose	
Early Wild	111
Pasture	111
Rosin Weed	34
Rough Blazing Star	26
Rough White Lettuce	29
Round-headed Bush Clover	59
Scribner's Panic Grass	94
Scurfy Pea	62
Sedge	
Bicknell's	45
Mead's	46
Oak	47
Pennsylvania	47
Prairie	45
Strict	48
Tussock	48
Seneca Snakeroot	102
Shooting Star	103
Silphiums	33
Showy Tick Trefoil	57
Side-oats Grama	88
Silky Aster	14
Sky-blue Aster	10
Sloughgrass	97
Smooth Beard Tongue	118
Smooth Blue Aster	12
Spiderwort	44
Stiff Goldenrod	40
Strict Sedge	48
Sunflower	20
Downy	20
False	23
Prairie	22
Western	21
Switch Grass	95
Tall Coreopsis	18
Tall Green Milkweed	6
Thimbleweed	105
Thistle	
Pasture	16
Prairie	16
Tick Trefoil	57
Canada	57
Illinois	58
Showy	57
Tiger Lily	75
Turk's Cap Lily	75
Turkey Foot	86
Tussock Sedge	48
Vanilla Grass	92
Violet Wood Sorrel	83
Western Sunflower	21
White Lady's Slipper	80
White Lettuce	
Glaucous	29
Rough	29
White Prairie Clover	60
White Wild Indigo	55
Wild Bergamot	69
Wild Hyacinth	73
Wild Quinine	28
Wild Strawberry	108
Wood Betony	117
Yellow Cone Flower	30
Yellow Star Grass	74
Yellowish Gentian	65